This Book

was presented to

Rotherham College of Arts and Technology

Date September 1998

Rotherham College of Arts & Technology
R00053

STEELMAKING

C. MOORE and R. I. MARSHALL

The Institute of Metals
1991

Book Number 460

Published in 1991 by
The Institute of Metals
1 Carlton House Terrace, London SWIY 5DB

and
The Institute of Metals
North American Publications Center
Old Post Road, Brookfield VT 05036
U S A

Compiled by the Institute's CRC unit from original typescript
and illustrations provided by the authors

British Library Cataloguing in Publication Data
Available on request

I S B N 0-901462-73-X

Library of Congress Cataloging in Publication Data
Applied for

Made and printed in Great Britain
by The Bourne Press Ltd, Bournermouth

Cover photograph: Charging the AOD with molten stainless steel at British Steel Stainless, Sheffield.

CONTENTS

PREFACE

This monograph has been prepared with the objective of informing undergraduates, metallurgists and others interested in steelmaking on the status of the various methods currently available throughout the world.

Since the monograph is written as a concise "stand alone" statement of steelmaking processes detailed text references have not been provided. Any reader requiring a comprehensive source of references on a particular aspect of a process would find such data in the Institute of Materials bibliographic tabulations.

It is a complete revision of the original Monograph 6. New chapters have been included and others have been re-written to reflect the changes in the steel industry which have occurred in the last decade.

ACKNOWLEDGEMENT

The authors would like to acknowledge the help from BOC Industrial Processes Department in both the preparation and checking of the text for this monograph, and the Public Relations Department of BOC Limited for the preparation and design of the cover.

CHAPTER 1
Development of the Iron and Steel Industry in Britain 2000 BC to 1856

Iron is the fourth most abundant element in the earth's crust and ore deposits containing substantially more than the average of 5% Fe are widely dispersed throughout the world. Iron is a reactive metal and is not found naturally in elemental form except for small quantities in, for example, meteorites. For the same reason it is difficult to say how long iron has been smelted since, as all motor car owners know, iron and steel have a tendency to revert to the hydrated oxide in years rather than in decades, let alone centuries. The oldest surviving iron artefacts probably date from about 2,000 BC but there is evidence that the metal was smelted before this date in China, India, Europe and other locations. No-one knows who first deliberately extracted iron from ore as the remains of similar, small, medieval furnaces, often situated in rocky hollows on the windward side of hilly sites have been found in many parts of the world.

Iron was extracted in Britain before the Roman occupation which started with the arrival of Anulus Plantius in 438 BC and lasted until 383 AD. During this 800 year period greatly improved methods of smelting were developed and in these natural draught was augmented by foot operated single acting bellows which were introduced by the invaders. The withdrawal of the Roman legions by Magnus Maximus in 383 AD to support an attempt to occupy NW Europe marked the beginning of a 1000 year recession in iron smelting in Britain which lasted until the reign of Edward III in the 14th century.

An alternative to the bellows method of creating a forced draught for smelting was utilised in the Catalan process which was developed in Spain to extract iron using charcoal as a fuel from rich Pyrenean ore. In the Catalan process a low pressure blast was created by entraining air with free falling water in an open ended pipe or "trompe". The Catalan process was adapted and improved in America where it led to the development of the American Bloomery and the Champlain Forge in which the air blast was raised to a pressure of 1.5 to 1.75 psi (lb/in^2) and a temperature of about 300°C before it was used to combust the charcoal mixed with ore on the furnace hearth. The Catalan and American Bloomery methods of making wrought iron were extremely inefficient in terms of fuel consumption.

It should be noted that the above processes produce wrought iron by the "direct" method, ie iron ore is directly produced using charcoal to produce a slaggy but ductile form of wrought iron. The soft, oxidised wrought iron could subsequently be converted into a stiffer form of "blister steel" by "cementation".

The cementation process was developed in the 17th century and involved treating batches of malleable iron bars in an externally heated and enclosed vessel for the best part of a week at temperatures ranging from 800°C to 1000°C in the presence of charcoal and various activators like cowhorn and hoof which were found to increase the diffusion rate of carbon into iron. Time and temperature determined the rate of carbon diffusion into the blister steel and the progress of carburisation was monitored by the periodic withdrawal and testing of "trial bars" until satisfactory properties were achieved. After cementation the blister steel was rolled, forged and hammer welded to homogenise the structure and to produce various grades which were classified as shear steel, double shear steel or spring-heat steel etc.

In 1740 Benjamin Huntsman, a Doncaster clockmaker, devised his "enclosed blacksmiths hearth", a crucible method of melting blister steel to make an alloy with better and more consistent properties than could be achieved by the solid phase forging and manipulation of blister steel. The Huntsman Crucible process was widely used in South Yorkshire and the high reputation that Sheffield Steel enjoys throughout the world was contributed to by the process.

It is interesting to note that although "direct" ironmaking methods were used in this country until the 18th century they are often thought of today as a new concept. At present metallurgical coke is still available and is the most economic reducing medium and fuel for ironmaking. It may be however, that at some time in the future, fossil fuels will have to be more carefully used and although reducing gases might be prepared from them to make sponge iron directly from ore, the energy for smelting will most probably be generated in nuclear power stations. Almost all the steel used in the world today is made "indirectly" by oxidising carbon and impurities out of "hot metal" or molten pig iron from the blast furnace.

Blast furnaces use carbon as a fuel and reducing agent and since the conditions inside the furnace are strongly reducing the resultant alloy is saturated with about 4% carbon. In most cases changing pig iron into steel involves the removal by oxidation, of more than 95% of the carbon present together with other impurity elements. Oxidation may take place in the solid pasty stage as in Cort's puddling process or by refining liquid iron.

Small shaft furnaces have been used to smelt iron ore since the earliest days of ironmaking but by the 16th century vast areas of afforested land were being laid bare to produce charcoal for ironmaking. The thermal efficiency of a blast furnace is closely related to its working volume. Charcoal is friable and is not strong enough to sustain a high column of iron ore in a large volume furnace and its low density leads to high dust losses at high blowing rates.

For the above reasons it was evident that a stronger and less wasteful fuel than charcoal would have to be utilised in the blast furnace and in 1620 Dud Dudley succeeded in smelting iron with coal but unfortunately his process was kept secret and died with him in 1684. It was not until 1709 that Abraham Darby succeeded in operating the Coalbrookdale blast furnace on coke and during the succeeding 50 years coke completely superseded charcoal in British blast furnaces and ironworks were gradually relocated in the coalfields and away from the woodland areas of the country. The next major improvement in blast furnace technology did not take place until 1828 when James Neilson introduced the hot blast (300°C) at Clyde Ironworks in Glasgow.

Unfortunately 4% carbon pig iron is not malleable enough to forge so that Henry Cort's puddling process, which was patented on 13 February 1784, was a tremendous breakthrough. Cort's dry puddling process involved the manipulation of a pasty slag-metal mixture on the hearth of a reverberatory furnace. Puddling was a backbreaking job and in addition to the severe physical demands the operators had to work in a very hot and unpleasant environment. After thoroughly stirring up the metal in the furnace to encourage carbon and silicon removal, the pasty mass was manhandled from the puddling furnace in 40 kg lumps, consolidated and homogenised by passing through "a grooved roll mill" patented by Cort in 1783.

Britain's emergence as the major industrial country of the world in the 19th century resulted from the efforts of the plethora of brilliant innovators and engineers who lived and worked between 1750 and 1850. Many of these men of genius made contributions in more than one field but the steel industry of today owes a particular debt not only to intellectual giants like Bessemer, Siemens, Darby, Cort, Murdock and Nasmyth for their metallurgical contributions but also to people like Brunel, Smeaton and Telford where insight allowed them to make demands which stretched the innovators.

During the same period Matthew Boulton and James Watt developed and perfected the condensing steam engine and Watt of course defined one horsepower as the work done in raising 550 lb through 1 ft in 1 s. The importance of the invention of the steam engine to the emergent steel industry of Victorian times cannot be over emphasised. Boulton and Watt steam engines, replaced water wheels as an improved power source for blast furnace blowing engines, forges, hammers and rolling mills.

In 1825 George Stephenson successfully applied steam power to transport passengers on the Stockton to Carlisle railway and the success of this venture lead to the rapid development of an extensive railway network in Britain for carrying goods and passengers. Setting up the railway system benefitted the industry by creating a large demand for permanent way materials and rolling stock. During the 19th century steel gradually superseded wood as the material used to build large ships and by the turn of the century steam ships which could sail into wind were replacing sailing boats.

Isambard Kingdom Brunel (1806-1859) was a great civil and railway engineer of the period and in addition to building the Clifton suspension bridge over the River Avon in Bristol he designed and built the ocean liners Great Western and Great

Britain. During an early design stage of the Great Britain which was to be an iron ship displacing about 3000 tons, Brunel considered the possibility of powering the ship with paddle wheels. This plan was later abandoned and the ship was eventually driven by a screw propeller.

While investigating the possibility of using paddle wheels on the future transatlantic liner Brunel approached one of the most colourful, accomplished, and seldom acclaimed innovators of the age, ie James "Steam Hammer" Nasmyth, to devise a method for forging the great paddle shaft for his 100 m long ship.

Nasmyth was born in Edinburgh in 1808 and came from celebrated Scottish stock. One of his ancestors, Elsbeth Naesmith, was burnt at the stake for witchcraft and his father was the Edinburgh artist and architect Alexander Nasmyth, who was a friend of Sir Walter Scott and who painted the "definitive portrait" of Robert Burns. Brunel's request led Nasmyth to design a steam hammer in 1839 in which the hammer was connected to a steam piston which raised the hammer and allowed it to fall under gravity onto the workplace. Nasmyth subsequently improved the design to allow the hammer to be propelled downwards under steam pressure to strike a "dead and non-bouncing" high impact blow.

The hammer's descent could be very accurately controlled by steam pressure and one of Nasmyth's favourite demonstrations was to crack an egg in a wineglass with his 5 ton hammer without breaking the glass. Nasmyth was contracted by the Admiralty to supply forging hammers to the Royal Dockyards to enable heavy anchors and other items to be forged. He subsequently produced a hammer which was used for piledrving work in the construction of Naval Dockyards. Apart from inventing the steam hammer, Nasmyth's other metallurgical claim to fame is that in 1854 he patented a method for "wet puddling" and making cast iron malleable by agitating molten iron with a blast of steam. The process worked very well and was much less physically demanding than Henry Cort's puddling process.

In 1855 Henry Bessemer obtained his patent for "The Manufacture of Iron and Steel without Fuel" and the process became established in 1856. Bessemer recognised the similarity between Nasmyth's process of "wet puddling" and his own invention and offered Nasmyth a one third share in the royalties from his patent. Nasmyth declined the offer as he was by this time affluent enough to retire and he moved down to Kent where he died in 1890 at the age of 82 after achieving distinction as an astronomer.

CHAPTER 2
Review of Now Redundant Steelmaking Processes operated between 1850 and 1970

BESSEMER AND THOMAS CONVERTERS 1856-1970

Sir Henry Bessemer invented his process for the conversion of low sulphur and phosphorus haematite iron, into steel, in 1856.

Bessemer produced malleable steel by blowing cold air through liquid pig iron or hot metal and this resulted in the exothermic oxidation of carbon, silicon and manganese.

The original converter was operated with an acidic silica refractory lining and an acid slag. The significance of slag basicity was not appreciated in the early days of steelmaking and even Bessemer was frustrated in his efforts to remove phosphorus from iron in his acid lined vessel. A diagram of a Bessemer Converter in operation is shown in figure 2.1.

In 1878, Sidney Gilchrist Thomas, a London Police Court clerk and his cousin Percy Gilchrist, who was employed as a works chemist in a South Wales ironworks, developed a basic lined version of the Bessemer converter in which a dolomite lining of calcined $CaCO_3$ $MgCO_3$ was used.

Gilchrist and Thomas found that if steelmaking was carried out in their basic lined vessel, with a basic lime slag, phosphorus, and some sulphur could be removed from the metal during steelmaking to form calcium phosphate and sulphide compounds in the slag.

The development of the Basic Bessemer or Thomas Converter process allowed the Minette phosphoric limonite orefields of

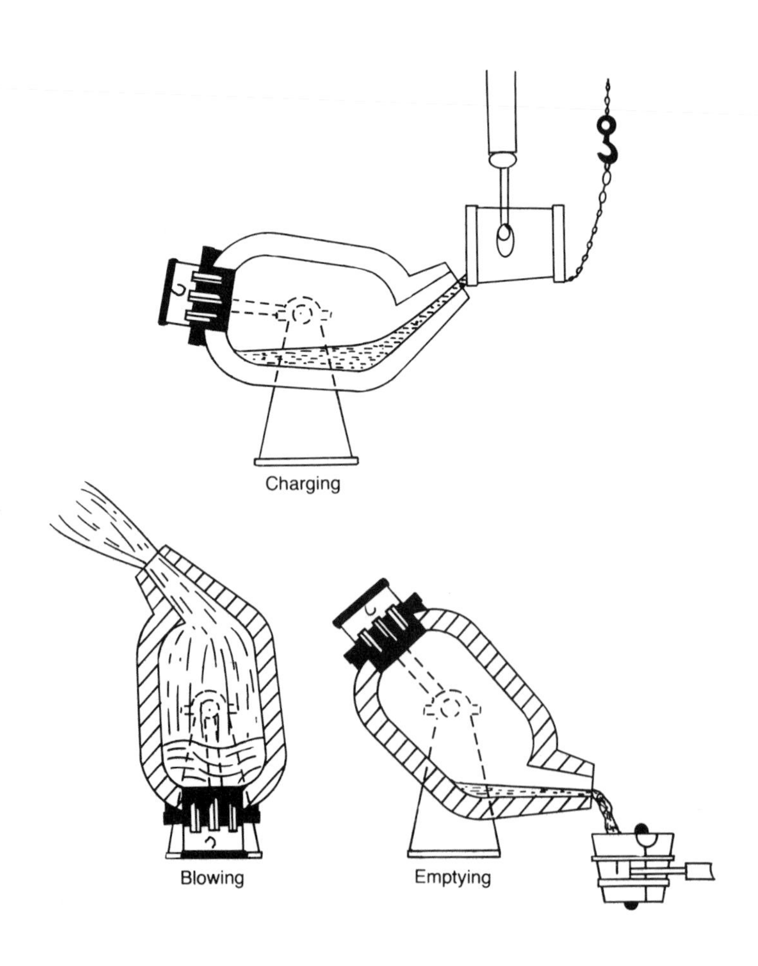

2.1 Bessemer converter operation (Bashforth)

France and Germany to be exploited and the process enjoyed the status of being the major steelmaking method employed in Europe until it was superseded in the 1960s by top blown oxygen converter methods of steelmaking.

Features of Bessemer and Thomas Converter/Steelmaking

The above processes are identical in concept but, as Table 2.1 shows, the compositions of the hot metal used were quite different.

In the acid Bessemer most of the heat evolved came from the oxidation of silicon whereas the basic Thomas converter derives most of its energy from phosphorus oxidation.

Figures 2.2a and 2.2b show characteristic refining curves for both processes.

Table 2.1

Iron analyses for Bessemer and Thomas Iron

Element %	Bessemer Iron	Thomas Iron
Carbon	3.5-4.2	3.3-4.2
Silicon	2.0-2.5	0.3-0.6
Manganese	0.5-1.5	0.6-1.1
Phosphorus	0.04 max	1.8-2.0
Sulphur	0.04 max	0.05-0.07

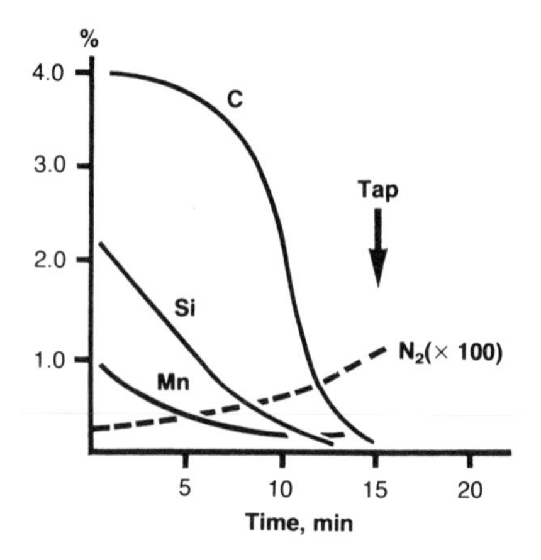

a) Acid Bessemer Converter

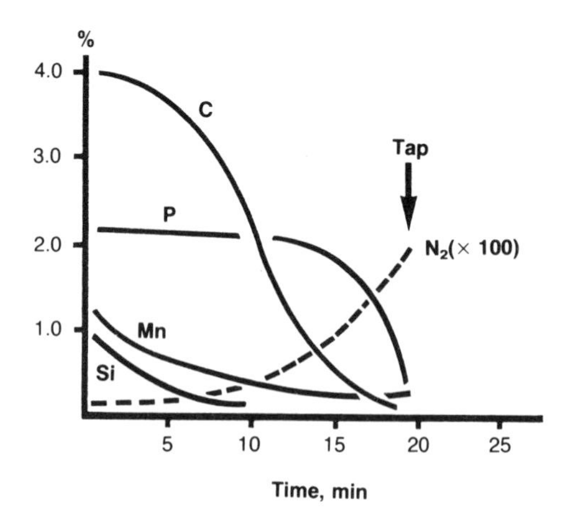

b) Basic Bessemer Converter

2.2 Metal Composition Changes during blowing

Figures 2a and 2b show that:

1. Carbon removal, in both acid and basic processes, was rapid.

2. Silicon and manganese were oxidised out early in each blow.

3. Phosphorus was only removed in the Basic Bessemer during the "afterblow" period, due to the higher slag iron when the carbon content is low. In the BOS process, which will be considered later, phosphorus is removed during decarburisation.

4. Sulphur was not removed during acid Bessemer steelmaking but up to 30% could be removed in the Thomas process by operating the furnace with a basic slag.

5. The exothermic oxidation of impurities resulted in a progressive temperature increase from 1300°C to 1650°C. Because of the large amount of heat carried out of the vessel by nitrogen neither converter was able to consume more than a small percentage (10% max) of scrap steel which was generally added as a coolant.

6. The nitrogen content of air blown converter steel built up to a high level of between 0.015 and 0.020%, especially during the latter part of the blow when the rate of carbon monoxide evolution was minimal.

Rapid analytical methods were not available during the lifetime of Bessemer and Thomas converters, contemporary methods were able to evaluate carbon in a minimum of 6 min and manganese in 20 min.

Because of the very rapid carbon drop rate achieved compared to the Open Hearth process, in practice "catch carbon" was not possible and little control was exercised during the blow. Refining was judged to be complete shortly after the carbon monoxide flame dropped in the mouth of the vessel. In the Thomas Converter blowing was slightly prolonged after carbon removal so that phosphorus was actually oxidised out during a period known as the "afterblow". In either case excessive blowing resulted in high yield losses "dirty steel" due to iron oxide formation and a high nitrogen content.

Post-War Developments in Bessemer Steelmaking - the VLN Process

The availability of tonnage oxygen during the immediate post-war years lead to the development of the LD process (and variations) and the Kaldo and Rotor processes. During the same period attempts were made to "pep-up" the performance of Open Hearth and Bessemer furnaces by the oxygen enrichment of combustion and blown air.

In the case of the Bessemer converter it was found that even a modest oxygen enhancement of the air blast lead to unacceptably rapid bottom wear resulting from increased reaction temperatures. Successful steelmaking was, however, carried out in bottom blown converters using a coolant gas with oxygen. Carbon dioxide and steam were used for this purpose but Carbon dioxide was not used to any great extent because of its high cost.

The Basic Bessemer furnaces of Port Talbot were, however, successfully modified to produce VLN (very low nitrogen) steel during the 1950 s and 1960 s using injected oxygen-steam mixtures. VLN steel was particularly suitable for rolling into deep drawing grades of non ageing strip for the motor car industry as the thinness of the final product allowed hydrogen to diffuse freely out of solution without causing problems.

Conclusion

During their useful lifespan of 100 years Bessemer and Thomas Converters were used throughout the world and were the mainstay of the Continental European steel industry.

The main disadvantages of the processes were:

1. The high nitrogen content of the steel produced.

2. Their inability to consume more than 5-10% of cold scrap.

Sir Henry Bessemer was unquestionably the greatest iron- and steelmaking genius of all time and his converter re-emerged in modified form as the Q-BOP process which is discussed in Chapter 4.

THE OPEN HEARTH STEELMAKING PROCESS FROM 1865 ONWARDS

Until the 1950s the Open Hearth process was the most important steelmaking method used in Great Britain and the United States of America and accounted for over 90% of the steel made in these countries at that time. The process enjoyed less popularity in continental Europe where the Thomas Converter or Basic Bessemer process was better suited for dealing with the high phosphorus hot metal made from indigenous ores. The process has virtually disappeared from the West but as recently as 1988 over 96 million tonnes of Open Hearth steel was made by Eastern Bloc countries and this represented 46.2% of their total output.

The Open Hearth process dates from 1865 when Sir William Siemens first made crucible steel in a regenerative furnace in Birmingham while simultaneously in France Pierre Marten was developing a similar method for melting pig iron/steel scrap mixtures. Open Hearth steelmaking reached its zenith in the 1950s during which time oxygen assisted, oil fired, all basic furnaces achieved production rates of up to 20 tonnes/hour. The process has been extensively investigated over the years and our understanding of steelmaking reactions owes some debt to the generations of research workers who have been involved with Open Hearth steelmaking. The process is instructive to metallurgists since it involved the basic thermodynamics and slag-metal chemistry to obtain the right temperature and compositional requirements for tapping. A diagram of a gas fired open hearth furnace of traditional design is shown in figure 2.3. The furnaces could be fixed on tilting.

Open hearth furnaces are the most versatile of all steelmaking units in that they can accept metal charges ranging from 100% liquid to 100% solid. Open hearth furnaces may make steel from 100% blast furnace hot metal but mixtures of cold scrap steel and molten or cold pig iron have generally proved more economic in the United Kingdom. Depending on availability and cost, a conventionally operated open hearth furnace usually consumes between 30 and 70% of cold steel scrap.

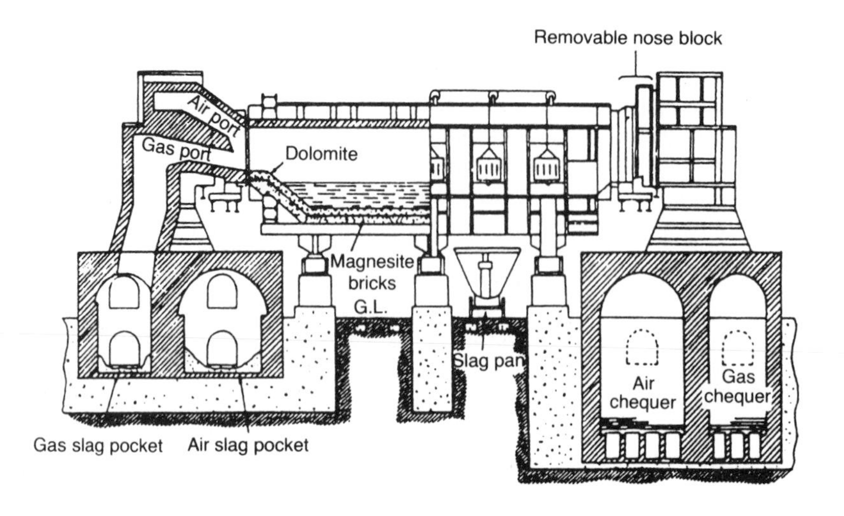

2.3 Section through a 200t fixed open hearth furnace (Jackson)

Open Hearth Steelmaking

The open hearth steelmaking process can be divided into four stages:

1. Charging
2. Melting
3. Refining
4. Fettling

Charging

Open hearth furnaces are charged through water cooled doors set in the front wall of the furnace. In basic furnaces scrap steel is usually charged before the more acidic pig iron or hot metal which would otherwise attack the furnace hearth lining. The limiting charging rate is often governed by the melting and fuel input rates to a furnace but also relates to the percentage of liquid metal in the charge.

Melting

During charging, and the early stages of melting, surface oxidation of steel scrap occurs and this is supplemented by additions of cinder, millscale and iron ore, to help promote a vigorous carbon monoxide boil after liquefaction of the charge. The time taken to "clear melt" a charge depends on the proportion of "cold metal" used, the fuel type and input rate, and whether or not oxygen enrichment of combustion air is used.

Refining

Although some elements are reduced during charging and melting, the refining stage of steelmaking starts after melting of the charge is complete. Refining is particularly concerned with the removal of carbon from the iron by oxidation to a level determined by the specification. During refining sulphur and phosphorus are also removed from the charge by partitioning these elements between slag and metal in basic lined furnaces.

Because of their high affinities for oxygen, silicon and manganese are rapidly reduced to equilibrium level during refining. A typical "refining plot" for an open hearth charge is shown in figure 2.4.

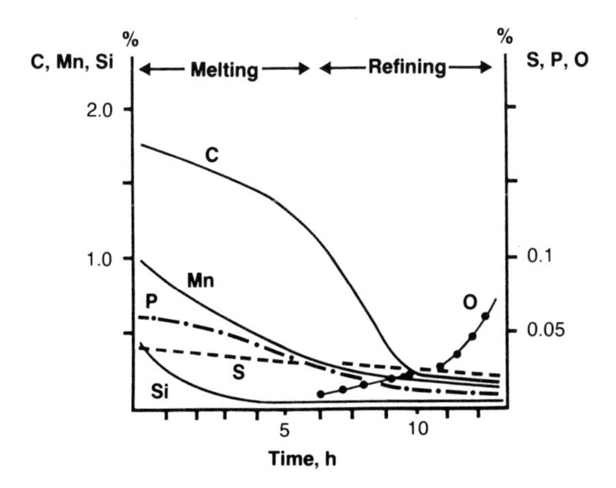

2.4 Composition changes during refining in the open hearth of a 50% hot metal-scrap charge

Fettling

After tapping, furnace hearths are repaired or "fettled" by splashing out slag pools and filling holes and worn banks with dolomite in basic furnaces, or silica sand, in the less commonly used and smaller acid furnaces. Fettling takes place as soon as the furnace is tapped while the lining is very hot and refractory patching material is fritted onto the damaged and worn parts of the furnace.

Steels made by the Open Hearth Process

The process can be used to make rimming, balanced, killed and low alloy steels containing up to 5% of alloying elements.

Conclusion

The Open Hearth Process is 125 years old and although declining in popularity it was used to make almost 110 million tonnes of steel or 16.1% of the worlds output in 1988. It is unlikely however that the method will survive into the 21st century. Open Hearth furnaces are labour intensive, complex in design, costly to build and because of their large wall and roof areas, have a low thermal efficiency. Even the advanced Ajax furnaces of Scunthorpe and the currently used Maerz-Bolenz units were, and are, unable to make steel at more than 35 t/h which compares unfavourably with the 500 t/h available from a modern BOS vessel.

Open Hearth furnaces were the first major gas fired units which by using regeneration to preheat gas and air allowed large tonnages of steel to be made from a 100% cold charge. The versatility of the process in terms of its charge flexibility and range of qualities of steel which could be made was particularly useful by this country during the Second World War.

THE KALDO PROCESS 1948-1970

The Kàldo process was developed, between 1948 and 1954, by Professor Bo Kalling at the Domnarvet Works of Stora-Kopparberg AB in Sweden. The location of the development is slightly surprising as the Domnarvet Works generates its own hydroelectric power and is situated near extensive magnetite deposits which would seem to favour an arc furnace method using pelletised stock. Essentially the Kaldo furnace was a rotating converter tilted on its side during operation as shown in figure 2.5.

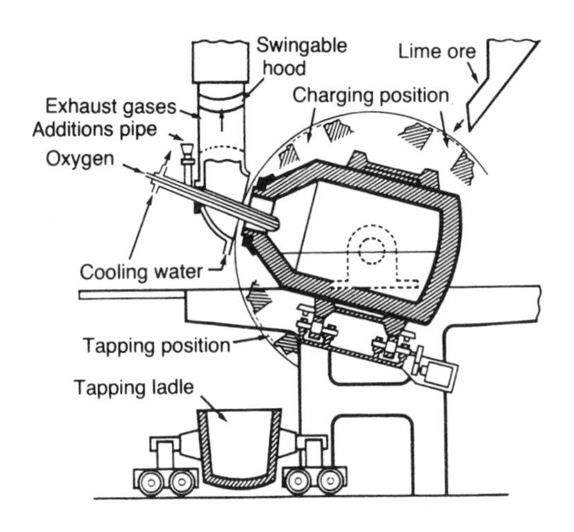

2.5 The Kaldo vessel

Kaldo vessels were designed to operate with the central axis of the vessel inclined at an angle of between 16° and 20° to the horizontal. Rotation at speeds of up to 40 rpm resulted in hot metal being carried up the vessel wall and then turned over on itself giving intimate mixing of slag and metal. The water cooled oxygen lance was designed to oscillate and the angle of inclination to the metal surface could be varied between 20° and 30° to the metal surface. The above arrangement altered the proportions of oxygen needed for refining reactions in the molten charge and for the combustion of carbon monoxide within the vessel to be controlled.

Oxygen was injected at approximately the same ratio as into BOF converters, ie at approximately 50-55 Nm^3/t but mechanical mixing allowed the utilisation of a softer blowing (3 bar) oxygen lance. The process was particularly useful for refining high phosphorous hot metal as slag/metal reactions rather than direct oxygen/metal refining was favoured by the agitation associated with rotation of the vessel. During steelmaking a Kaldo vessel tapping 100 tonnes of steel would have a tap to tap time of about 1.5 tonnes and an oxygen blowing time of about 1 hour.

Although the production rate of a Kaldo was lower than that of a BOF vessel of equivalent size, the better thermal efficiency resulting from heat being recovered from carbon monoxide combustion within the vessel allowed up to 50% cold scrap to be utilised in the charge compared to 30% in the LD. Unfortunately enhanced scrap consumption led to erosion and abrasion during operation and this resulted in rapid lining wear although with good practice lining lives of 90-130 heats were achieved. Vessel rotation presented difficult and expensive engineering problems since at 40 rpm the periphery of a 100 tonne Kaldo vessel had a linear speed of about 50km/h. The rotating mass weighed about 300 tonnes and this led to considerable stresses on the drive mechanism.

Conclusion

The Kaldo process enjoyed limited success between 1960 and 1970 with production plants being operated in Sweden, England, France and the United States of America. Annual world capability never exceeded 10 million tonnes.

Although the Kaldo vessel was thermally more efficient than the BOF converter, energy savings were outweighed by the high capital and maintainance costs of the enormous drive motors and refractories. In retrospect, it is interesting to observe that within months of installation at Consett Works in the mid 1960s a Kaldo vessel was operated in the vertical, non-rotating, mode and so was used in effect as a very expensive BOF converter.

THE OBERHAUSEN ROTOR PROCESS 1948-1970

The rotor process was developed in Germany by Dr R Graef of Oberhausen at about the same time as the LD and Kaldo processes were evolving in Austria and Sweden.

The Oberhausen Rotor was perhaps the most unwieldy steelmaking unit ever devised as figure 2.6 shows. A typical oil fired 65 tonne capacity unit had an overall length of 14m, an internal diameter of 3.5m, and was engineered to rotate on its longditudal axis and tilt through 90°.

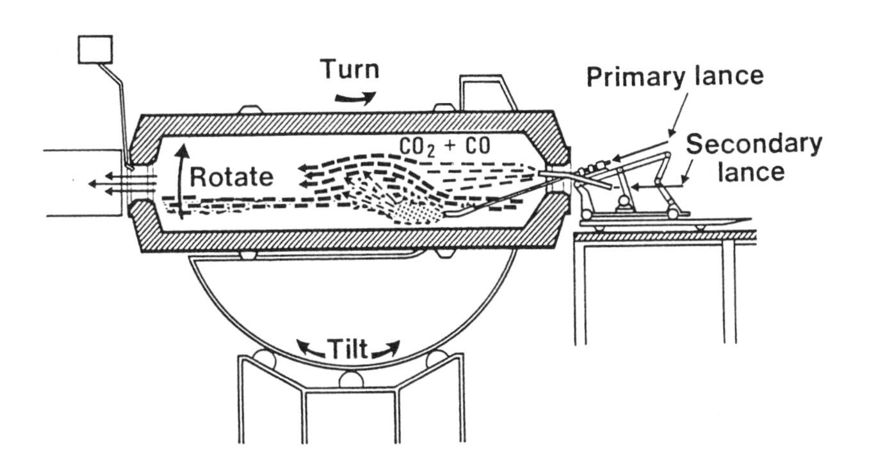

2.6 The Oberhausen Rotor vessel

The furnace consisted essentially of a long cylindrical vessel which rotated about its central axis at 2 to 4 rpm. Rotation resulted in good slag-metal mixing and oxygen was injected through submerged and secondary lances. Lance movement relative to the cylinder allowed refractory wear to be evened out and vessels were designed to be reversible to further spread wear.

The Oberhausen Rotor was devised to prerefine or make steel from high phosphorus hot metal containing up to 2.0% phosphorus. Scrap utilisation was not high and when added scrap was charged through small doors at either end of the vessel. Metallurgical reactions were similar to those obtained in LD-AC practice and because of the low rate of rotation of the vessel, ie up to 4 rpm compared to 30 rpm, lining lives were achieved which were considerably better than those obtained in Kaldo vessels.

A 100 tonne vessel could produce approximately 1000 tonnes of steel per day at production rates of 30-70t/h. The thermal efficiency of the process was improved as secondary oxygen or air allowed the combustion of carbon monoxide within the vessel and the heat evolved was transferred to slag and metal via the refractory lining. A few units were built in Germany, South Africa and the United Kingdom, but as far as is known, none is in current production.

Conclusion

The Oberhausen Rotor furnace was developed in Germany during the immediate post-war years and in common with developments taking place concurrently in other countries the method was conceived to take advantage of the availability of tonnage oxygen, distilled from the atmosphere, which became available for the first time during this period. The Rotor process was, like the Kaldo, clever in concept but its success was restricted by high installation and maintenance costs and its essential impracticability.

CHAPTER 3
The Chemistry of Steelmaking from Blast Furnace to Tundish

INTRODUCTION

The Concise Oxford Dictionary defines steel as a malleable alloy of iron and carbon much used as a material for tools, weapons, etc. Commercial steels are, of course, not simple binary alloys of iron and carbon as the above definition suggests since many elements are likely to be present and these include silicon, sulphur, phosphorus, manganese, nickel, chromium, molybdenum, copper and tin. Some of these elements are added in controlled amounts to enhance the properties of the alloy for a particular application whereas others originate from the scrap, fluxes and fuel used in steelmaking. Individual elements may be beneficial or detrimental depending on circumstances. For example chromium is a highly desirable ingredient in corrosion resistant austenitic stainless, or in 1% carbon 1% chromium abrasion resistant ball bearing steels but in boiler plate or strip steel the element would adversely affect weldability and ductility.

Although a small annual tonnage of steel is produced by electrically smelting gaseously reduced sponge iron most is made by oxidising and refining mixtures of molten pig iron (ie "hot metal") from the blast furnace and steel scrap in basic oxygen steelmaking (BOS) converters and electric arc furnaces (EAF's). BOS converters are alternatively referred to as BOF, ie basic oxygen furnaces.

Until 20 years ago steelmaking essentially took place in a furnace and deoxidising and alloy additions were made to the ladle on tapping. Customer specifications are now much more precise and analytical tolerances have been tightened so that techniques have become more sophisticated.

It has become increasingly common to treat liquid metal both before and after furnace refining takes place so that modern steelmaking can be divided into 4 stages as follows:

1. Hot metal desulphurisation (not in EAF practice)
2. Steelmaking
3. Deoxidation
4. Secondary steelmaking

High grade BOF steel might be subjected to all of the 4 stages listed above whereas steel for less exacting applications might simply be furnace refined and deoxidised. Three-quarters of the steel made in the United Kingdom is made by the BOS process which includes 70-85% of hot metal in the charge.

The refining reactions which take place during steelmaking are essentially the same for all processes and in the following section steelmaking chemistry will be discussed in general terms. Most of the impurities which have to be refined out during steelmaking come from hot metal so that it is appropriate to include a short section on ironmaking in this monograph.

IRONMAKING

Iron Ore

The four most abundant elements in the earth's crust are oxygen 47%, silicon 28%, aluminium 8% and iron 5%. Iron is found in oxide and sulphide form but since sulphur adversely affects the properties of steel, iron is invariably smelted from oxide ores. Iron is a moderately reactive element and combines with oxygen to form three oxides: FeO or wüstite, Fe_3O_4 or magnetite and Fe_2O_3 or hematite. Wüstite is unstable at temperatures below 570°C and iron is generally made from the most abundant, red coloured oxide hematite or from the black, rich, magnetic ore magnetite.

Commercial grades of magnetite and hematite contain between 50 and 65 wt-% iron and ores from different locations are usually blended, crushed and sintered to provide the blast furnace with a consistent burden material which would be at the higher end of the iron content range.

Blast Furnace Slag

The earthy waste material, or gangue, associated with ore and the ash from the coke charged and the coal injected to the blast furnace, is mainly a mixture of SiO_2 and Al_2O_3 and these acid oxides are fluxed with CaO to form the $CaO-SiO_2-Al_2O_3$ ternary slag system shown in figure 3.1 which is encountered in blast furnace smelting. The above

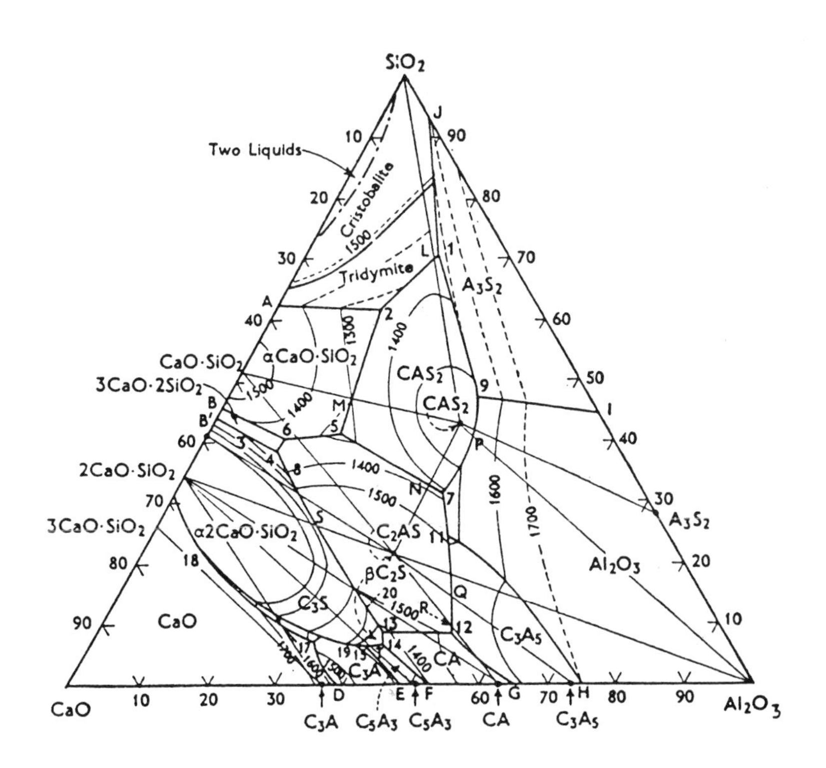

3.1 The $CaO-SiO_2-Al_2O_3$ slag system (Greig, Rankin and Wright)

three oxides have high melting points ie 2570°C, 1713°C and 2017°C respectively and so the composition of blast furnace slag has to be carefully controlled to keep it fluid and free running at a temperature which is consistent with a hot metal tapping at about 1500°C. Blast furnace slag can also contain appreciable amounts of sulphur (up to 3%) and when this is not fully removed from the iron (particularly plate iron) it adds significantly to the sulphur burden of the steelmaking unit.

Blast furnace slag is quite different from steelmaking slag in that it contains no iron oxide or phosphorus but will have very small amounts of iron physically entrapped. The material has commercial value as railway and motorway ballast and for brickmaking.

The Blast Furnace Process

A diagram of a modern blast furnace is shown in figure 3.2.

Although blast furnaces look much the same as they did 25 years ago changes have taken place which have resulted in increased productivity. Some of these might be summarised as follows:

1. Most modern blast furnaces operate on a burden which includes at least 80-90% of sintered or pelletised charge material.

2. Productivity has increased to such an extent that the Redcar 14m diameter hearth furnace, with a working volume of 3600m³, can produce 10,000 tonnes of iron per day.

3. Coke rates have dropped from 1,000 kg to less than 500 kg/tonne of iron produced during the last 25 years. Coal and oil injection at tuyere level has helped to optimise fuel consumption as have higher iron content ores with less gangue.

4. Conveyer belt charged, bell-less tops of the type pioneered by Paul Würth of Luxembourg are incorporated in all newly built furnaces. Most large, existing furnaces are being modified to the new top design.

5. Modern furnaces operate at high top pressure and campaigns have increased from 5 to an anticipated 10-15 years.

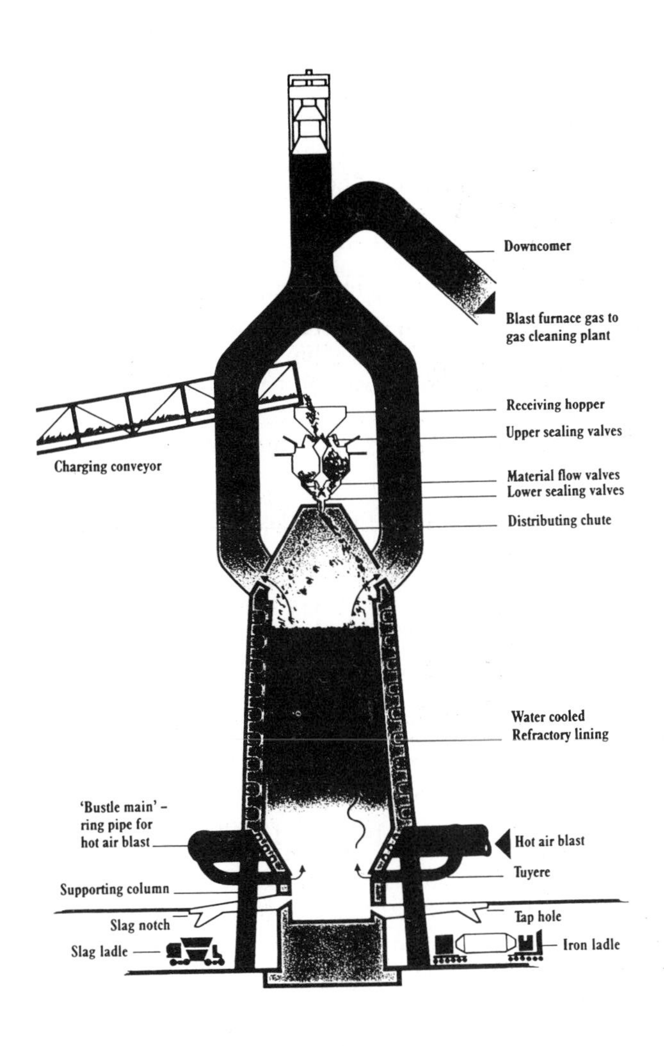

3.2 Profile of a Modern "bell-less top" blast furnace. (British Steel - information Services)

Blast Furnace Chemistry

Blast furnaces are charged with a mixture of (a) sintered or pelletised iron ore, (b) coke and (c) limestone.

During smelting these constituents change as follows:

(a) Sintered or Pelletised Iron Ore

Iron oxide may be considered to be progressively reduced during ironmaking, ie

$$Fe_2O_3 = Fe_3O_4 = FeO = Fe.$$

During reduction the initial iron content of the sinter or pellets which is usually in the range 55-65% of Fe by weight is increased to about 94% Fe in the liquid pig iron or hot metal produced. Since hot metal is in intimate contact with coke and ore in the blast furnace the iron produced is almost saturated with carbon and other solute elements are present as shown in Table 3.1.

Table 3.1 Typical Composition of Hot Metal*

Carbon	Silicon	Sulphur	Phosphorus	Manganese
4.5-4.7%	0.3-0.8%	0.02-.06%	0.06-.08%	0.3-0.8%

*Tapping temperature range 1450°C-1525°C

(b) Coke and Carbon

Coke serves two functions in the blast furnace, ie it acts as a fuel and as a reducing agent. It may be supplemented by coal and heavy fuel oil injection at the tuyeres to the extent of 25% of the equivalent coke rate. Metallurgical coke is made from a blended mixture of high grade coking coal and has to incorporate consistent quality with good abrasion resistance and reactivity. It contains about 10% of ash which has to be fluxed with lime and even the best grades of metallurgical coke contain about 0.7% of sulphur.

Carbon reduces iron oxide "directly" and "indirectly".

$$FeO + C \rightarrow Fe + C \quad \text{Direct reduction} \quad (2)$$

$$FeO + CO \rightarrow Fe + CO_2 \quad \text{Indirect reduction} \quad (3)$$

Direct reduction takes place in the high temperature regions of the blast furnace whereas indirect reduction takes place in the stack. Reduction is controlled by the "Carbon Deposition" or "Boudouard Reaction"

$$2CO = CO_2 + C \quad (1)$$

which encourages the breakdown of CO to CO_2 and C at low temperatures in the presence of carbon and vice versa as shown in figure 3.3.

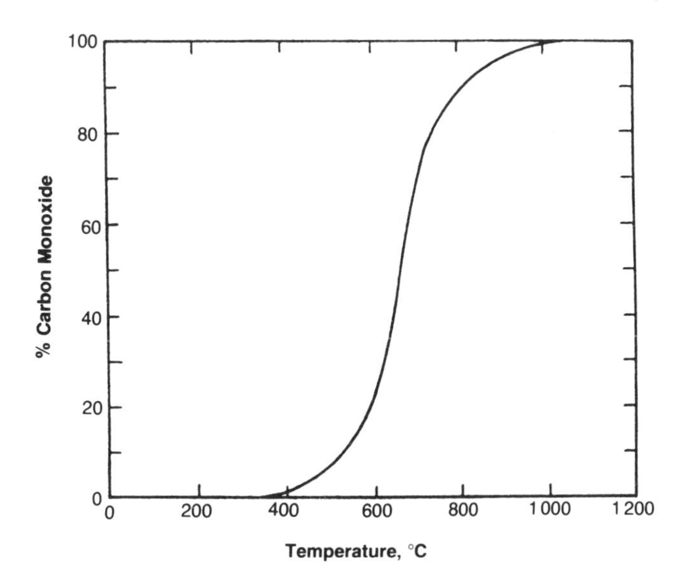

3.3 Variation with temperature of carbon monoxide in equilibrium with carbon dioxide and carbon at a pressure of one atmosphere.

If CO_2 were formed in the high temperature region of a blast furnace by the reaction:

$$2\,FeO + C \quad = \quad 2\,Fe + CO_2 \qquad (4)$$

The CO_2 would be unstable and by the Boudoard reaction would react with carbon as follows:

$$CO_2 + C \quad = \quad 2\,CO \qquad (5)$$

and by adding (4) and (5) we get

$$2\,FeO + 2\,C + CO_2 \quad = 2\,Fe + 2CO + CO_2 \qquad (6)$$

or

$$2\,Fe + 2C \quad = \quad Fe + 2\,CO$$

which is in effect DIRECT REDUCTION, ie equation (2) is the same as (6). Equation (1) is encouraged to go from left to right, by low temperatures and high pressure since it involves 2 volumes of CO being decreased to 1 volume of CO_2. Hence modern blast furnaces like the 10,000 tonnes/day furnace at Redcar operate at high top pressures of up to 2.2 bar.

The efficiency of operation of a blast furnace may be measured in terms of coke rate which should of course be as low as possible. The achievement of a satisfactory coke rate depends on optimising the extent to which the carbon deposition reaction proceeds. If the top gas is high in CO_2 sensible heat is carried from the furnace as a result of the exothermic reaction

$$2\,CO \quad = \quad CO_2 + C \qquad (1)$$

If on the other hand the top gas is high in CO, chemical heat leaves the furnace.

The final carbon content of hot metal depends on many factors including the time and temperature of contact between liquid metal and coke and also on the presence of solute elements which may increase the equilibrium carbon content as shown in figure 3.4.

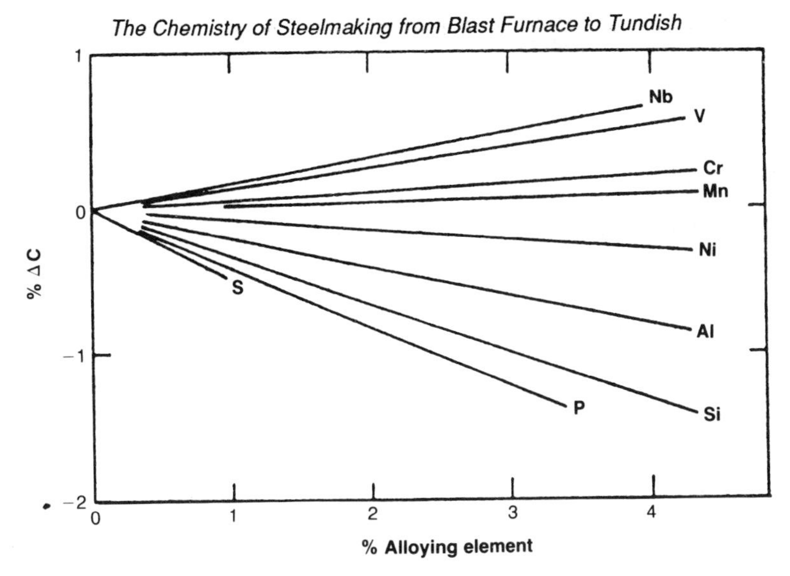

3.4 The effect of alloying elements on the solubility of carbon in liquid iron.

Silicon, Sulphur, Phosphorus & Manganese in Hot Metal

Silicon

The principal constituent of the gangue associated with iron ore and the residual ash from burnt coke is SiO_2, ie silica. In the strongly reducing conditions which prevail in the blast furnace thermodynamic considerations show that SiO_2 may be reduced by CO at temperatures in excess of 1490°C.

Silica is probably transferred from slag to metal in two stages, as equations (7) and (8) show. Please note that in this text rounded brackets are used to denote that the enclosed element or oxide may be considered to be dissolved in slag whereas square brackets imply solution in the metal.

$$(SiO_2) + CO \rightarrow SiO_{(gas)} + CO_2 \quad (7)$$

$$SiO_{(gas)} + CO \rightarrow [Si] + CO_2 \quad (8)$$

The above reactions take place in the high temperature bosh/hearth region of the furnace so that any CO_2 formed transforms to CO in the presence of carbon as indicated by equation (5) ie

$$CO_2 + C \rightarrow 2\,CO \quad (5)$$

by adding (7), (8) and (5) we get an overall reaction for silica reduction of:

$$(SiO_2) + 2C \rightarrow [Si] + 2CO \quad (9)$$

From the above consideration it can be seen that if a blast furnace runs "hot" this will probably result in the iron produced containing a high silicon content.

Because of the predominance of iron in the blast furnace under prevailing conditions some SiO_2 is reduced by iron as follows:

$$(SiO_2) + 2Fe \rightarrow (2\ FeO) + [Si] \quad (10)$$

The above reaction has some significance in sulphur partitioning and will be discussed in the next section.

Sulphur

Most of the sulphur which finds its way into the blast furnace is associated with coke. Good grades of metallurical coke might contain as little as 0.6-0.8% sulphur but sometimes coke with sulphur as high as 1% is used for ironmaking. Sulphur is liberated from coke on combustion and most is partitioned between slag and metal in the furnace. Blast furnace gas contains only a very small percentage of sulphur.

Desulphurisation is favoured by reducing conditions the reaction which takes place might be simply represented as follows:

$$[FeS] + (CaO) \rightarrow (FeO) + (CaS) \quad (11)$$

$$(FeO) + [C] \rightarrow Fe + CO \quad (12)$$

and by adding (11) and (12)

$$[FeS] + (CaO) + [C] \rightarrow (CaS) + Fe + CO \quad (13)$$

The above reaction shows that sulphur removal is favoured by high slag basicity and a high level of dissolved carbon in the metal which leads to a correspondingly low activity concentration of dissolved oxygen in the metal.

Sulphur and silicon contents are both related to the level of oxygen activity in the metal and for this reason it is difficult to achieve low levels of these elements simultaneously.

It has been shown that

$$(SiO_2) + 2\,Fe \quad = \quad (2\,FeO) + [Si] \qquad (10)$$

$$[Fe\,S] + (CaO) \quad = \quad (FeO) + (CaS) \qquad (11)$$

by the Law of Mass Action it can be seen that a high (FeO) content would push both of the above reactions from right to left and so favour a low silicon and a high residual sulphur content in the metal,

ie $[S] \propto \dfrac{1}{[Si]}$

Phosphorus

Although phosphorus imparts considerable fluidity to cast iron which enables thin walled castings to be made, it is an undesirable constituent in hot metal for basic oxygen steelmaking. Phosphorus pentoxide and iron oxide have very similar free energies of formation and since conditions in the blast furnace result in the reduction of more than 99.5% of the iron oxide charged, almost all the mineral phosphate which enters the furnace finishes up in the iron.

$$(P_2O_5) + 5C_{coke} \quad \rightarrow \quad [2P] + 5CO \qquad (14)$$

Phosphorus is reoxidised during steelmaking and most of the heat generated in the Basic Bessemer or Thomas converter process came from this strongly exothermic reaction.

The only way to produce low phosphorus hot metal from the blast furnace is to avoid charging burden materials which contain phosphorus.

Manganese

Most iron ores contain a little manganese oxide although some Canadian and North African ores might contain up to 2% Mn.

Manganese oxide is intermediate in stability between FeO and SiO_2, so that the oxide is reduced at temperatures above about 1370°C. Manganese partitions between slag and metal in the blast furnace and current British Steel practice commonly produces metal manganese contents in the range 0.3-0.8%. The value of having a high manganese content in hot metal is questionable as its high affinity for sulphur might make any ladle desulphurising treatment prior to steelmaking less effective.

HOT METAL PRETREATMENT - SULPHUR REMOVAL

Over the last 75 years many methods have been used to pretreat or partially prerefine hot metal before steelmaking to reduce sulphur, phosphorus, silicon and carbon. The main objective of all these treatments has however been to reduce the sulphur content of the hot metal and other elements have been removed almost incidentally. Sulphur could not be removed in acid steelmaking proceses like the Bessemer converter and the Acid Open Hearth Furnace and the high level of dissolved oxygen activity achieved in the modern BOF does favour desulphurisation. Equation (13) shows that sulphur removal is favoured by the high carbon content present in hot metal.

$$[FeS] + (CaO) + [C] \rightarrow (CaS) + Fe + CO \qquad (13)$$

Pretreatment has taken place in holding furnaces, torpedo ladles, but nowadays most treatment takes place in special ladles between the torpedo transfer ladle and the BOS vessel.

Ladle Desulphurisation

Many methods of desulphurisation have been and are being used. All methods, past and present, aim to remove sulphur from hot metal by adding or injecting materials which have a high affinity for sulphur, eg sodium carbonate, Na_2CO_3, lime, CaO calcium carbide, CaC_2, calcium cyanamide $Ca(CN)_2$, magnesium, Mg, etc. Agitation by mechanical stirring, vibration and gas bubbling helps to promote slag-metal reactions and in some cases, eg with calcium carbide injection or in the utilisation of coke breeze coated with powdered CaO an attempt is made to promote desulphurisation by increasing the carbon content of the metal as indicated by (13).

Soda Ash Treatment

Soda ash has been used to desulphurise hot metal for many years. The process was developed in conjunction with acid burdened blast furnace practice and the high sulphur hot metal so produced was treated by simply throwing bags of sodium carbonate into the tapping stream.

$$[FeS] + Na_2CO_3 \rightarrow (Na_2S) + CO_2 + (FeO) \qquad (14)$$

and under the high temperatures prevailing in the ladle

$$CO_2 + Fe \rightarrow (FeO) + CO \qquad (15)$$

also FeO combines with Mn and Si to form a mixed silicate slag

$$(FeO) + [Mn] + [Si] \rightarrow (FeO, MnO, SiO_2) \qquad (16)$$

Gas evolution in the ladle causes violent agitation of the liquid metal and when practised as described the process is likely to adversely affect the lives of ladle linings, and nearby workers, by alkaline oxide and fume attack. The process may be made less drastic by substituting a mixture consisting of 25% Na_2CO_3 and 25% CaF_2 and 50% $CaCO_3$ injecting the desulphurising mixture rather than adding it directly to the tapping stream considerably reduces fume evolution.

Magnesium Injection

Modern desulphurising treatments often involve a complex treatment involving a combination of a variation of the soda ash process in conjunction with magnesium granule injection.

Hot metal containing 0.06%S might have the sulphur level reduced to 0.03% by soda ash treatment and this can be further reduced to 0.01% or below by the injection of magnesium and lime granules using nitrogen as a carrier gas. Magnesium treatment usually takes place in an intermediate charging ladle between the topedo ladle and the BOS converter.

Different practices might involve the injection of any combination of desulphurising materials or might involve the injection of cored calcium wire if the additional cost is warranted by the quality of steel to be made.

Slag Removal

It is imperative that whichever of the above techniques is used all traces of sulphur rich slag is physically removed from the treatment ladle before the hot metal is changed to the BOF. If sulphur rich slag is not removed sulphur will revert to the metal when oxygen blowing commences.

$$(MgS) + (FeO) = (MgO) + [FeS] \qquad (17)$$

Mixing Furnaces

Large holding furnaces, called "mixers" sometimes with a capacity in excess of 500 tonnes used to be popular in integrated steel plants. These mixing furnaces received torpedo ladles full of hot metal from the blast furnace plant and allowed compositional differences in batches of metal and from different furnaces to be evened out and so provided a consistent composition of metal for the steelmaker. Mixer furnaces were usually gas fired and could be "inactive" holding furnaces or "active" furnaces in which a slag was made and a limited amount of prerefining took place.

STEELMAKING REACTIONS

Introduction

During ironmaking iron oxide, containing approximately 60% Fe by weight, is reduced to form pig iron or hot metal which as Table 3.1 shows consists of about 94% Fe plus impurities which are mainly carbon, silicon, sulphur, phosphorus and manganese. Steelmaking is an oxidising and refining process which involves the partial removal of the above elements to controlled levels determined by the specification of the steel being made. Plain carbon steel contains more than 98.5% and the process route from ore to steel may be represented by the following block diagram:

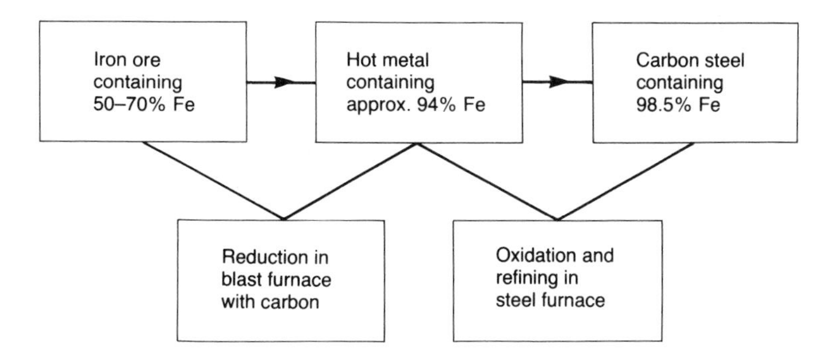

Steel is made by fluxing mixtures of hot metal and steel scrap with calcined limestone or CaO. Different steelmaking processes are characterised by the charges they accept. Electric arc furnaces can make steel from a 100% cold scrap charge whereas basic oxygen converters operate on mixtures of 70 or 80% hot metal with the balance being made up of cold scrap. The "metallurgical load" on a furnace is proportional to the hot metal charged. The economics of different steelmaking processes are directly related to the charges they accept and the relative costs of cold scrap and hot metal.

Elements Encountered in Steelmaking

Although steel is essentially an alloy of iron and carbon, alloying elements are always present and these may originate as:

(a) contaminants in the scrap melted

(b) elements present in the raw materials used, eg hot metal, fluxes and fuel

(c) gases absorbed during steelmaking

(d) elements added to meet the desired specification.

The above elements can in turn be divided into four sub-groups, ie:

1. Elements which can be almost completely eliminated during steelmaking.

2. Elements unaffected by steelmaking

3. The gases, oxygen, nitrogen and hydrogen

4. Carbon and elements which partition between slag and metal.

Elements Eliminated During Steelmaking Including Pb, Zn and Al

Lead is almost insoluble in liquid iron and it usually finds its way into steelmaking furnaces in plumbers scrap or as a constituent of solder or bearing alloys. Lead melts at low temperature and forms a fluid and high density liquid which seeps into furnace bottoms. During the days of the open hearth steelmaking "stalactites" of lead, often over a metre in length, were frequently found attached to the underside of 1.5 metre thick hearth linings.

Zinc usually comes from galvanised scrap and the metal melts and readily volatilizes at steelmaking temperatures. The vapour can penetrate the interstices of porous refractories and cause spalling. Zinc oxide, dust gathers in surprisingly large quantities and tends to choke the fume extraction system of electric arc furnace installations.

Aluminium is used extensively in motor car construction. All automobiles have aluminium alloy pistons, most have light alloy cylinder heads and some have cylinder blocks and gear box housings made from aluminium alloys. Inevitably some aluminium gets into steelmaking furnaces as a scrap constituent. At steelmaking temperatures aluminium oxidises readily and may increase the refractories of the slag rendering the slag less fluid and reducing its reactivity.

Summary

Pb, Zn and Al can be almost completely removed during steelmaking but all are undesirable since Pb may cause furnace bottom cracking, ZnO chokes the fume extraction system, and reactive metals like Al and Mg make slag refractory.

Elements Unaffected by Steelmaking Including Cu, Sn, Ni, Co, Mo and W

The above elements have less affinity for oxygen than iron so that they cannot be removed during refining. Although copper has a slightly beneficial effect on the corrosion resistant properites of steel both copper and tin are generally regarded as undesirable. Both elements cause hot shortness during rolling and tin embrittles steel alloys. Copper is generally restricted to 0.2% max and tin to 0.03% max in most grades of steel. Copper and tin get into steel by way of electrical and plumbing scrap, solder and tinplate.

Nickel, molybdenum, cobalt and tungsten are all highly desirable element in alloy and tool steels and enter steel scrap in these forms. The above elements harden and toughen iron and if present would make low carbon plain carbon steel grades unsuitable for deep drawing and welded applications.

Summary

Cu and Sn are usually undesirable in all grades of steel. Ni, Mo, Co and W are beneficial when present as specified elements in alloy and tool steels but they are not necessarily advantages in steel for more ductile applications.

The Gases - Oxygen, Nitrogen and Hydrogen

Steelmaking is an oxidising and refining process carried out in air. During refining, and if care is not taken during tapping and teeming the above gases are absorbed into and dissolved in liquid steel.

Oxygen

High oxygen levels lead to the formation of oxides - usually silica and alumina. These non metallic inclusions are particularly undesirable in quality steels since fatigue resistance and ductility are significantly reduced.

Nitrogen

During refining nitrogen is removed from hot metal by the purging action of carbon monoxide bubbles into which dissolved nitrogen diffuses. As bath carbon falls, and the intensity of carbon monoxide evolution decreases, bath temperature increases and this combination of circumstances results in a tendency for nitrogen pickup to occur towards the end of the blow.

Nitrogen content is related to the purity of oxygen blown during steelmaking.

Sievert's Law predicts that if equilibrium is achieved the solubility of nitrogen in the liquid steel will be in proportion to the square root of the partial pressure of nitrogen present.

i.e. $$N_{(in\ Fe)} = K\sqrt{P_{N_2}}$$

Because of the foregoing it is particularly important to use very pure oxygen especially for deep drawing quality steels which are susceptible to strain age embrittlement.

Oxygen purities of 99% or better give rise to steels containing less than 0.003% nitrogen which is low enough for most applications.

The final nitrogen content is also influenced by the nitrogen content of hot metal which can vary from 0.003 to 0.009%.

A major limitation of the Bessemer process was the high nitrogen pickup from the air blast. Bessemer steels are characterised by final nitrogen contents of 0.012 - 0.018% which makes them unsuitable for low carbon steels for deep drawing operations.

Hydrogen

Because of the low partial pressure of hydrogen during BOS the final hydrogen content of steel is seldom above 3 ml/100g. The process is therefore particularly suitable for making low and medium carbon steel for strip, plate and stacked steel plate applications. Such steels generally have a low crossectional thickness and are not susceptible to hair line crack formation and flaking at the low levels of hydrogen present.

On the other hand steels made by the electric arc furnace process often contain up to 10ml/100g or ppm of hydrogen and this makes low alloy steels of moderate to heavy section especially prone to hairline cracking.

In the electric arc furnace moisture from the atmosphere is dissociated by the arc which also converts molecular hydrogen into very small and mobile monatomic hydrogen atoms which diffuse readily into the alloy and subsequently cause problems with cracking. Water vapour is dissociated by the arc to oxygen and molecular hydrogen

$$H_2O \rightarrow \frac{1}{2} O_2 + H_2$$

Molecular hydrogen converts to atomic hydrogen in the arc

$$H_2 = 2H$$

Vacuum degassing plant is available in most modern plants and although vacuum treatment can remove 90% of the hydrogen present in liquid steel it cannot remove more than 10% of the nitrogen.

Summary

Oxygen, nitrogen and hydrogen are undesirable in steel. Vacuum degassing techniques can be employed to remove most of the hydrogen and some of the oxygen but are ineffective against nitrogen. For high grade applications liquid steel streams should be shrouded from the atmosphere to minimise gas absorption.

Carbon and Elements Which Partition Between Slag and Metal Including Si, S, P and Mn

Steelmaking is principally concerned with the control of the above elements to within prescribed limits. Table 3.2 shows the corresponding compositions of hot metal and a low carbon refined steel before alloying additions have been made.

Table 3.2 Composition of hot metal and steel

Composition	Carbon	Silicon	Sulphur	Phosphorus	Manganese
Hot Metal	4.6	0.65	0.04	0.07	0.55
Steel *	0.05	trace	0.015	0.015	0.15

*To allow for subsequent ladle treatment and continuous casting such a steel would probably be tapped at 1670°C-1690°C.

Fig 3.5 shows the rate of removal of the elements contained in the above table during the course of a BOS blow.

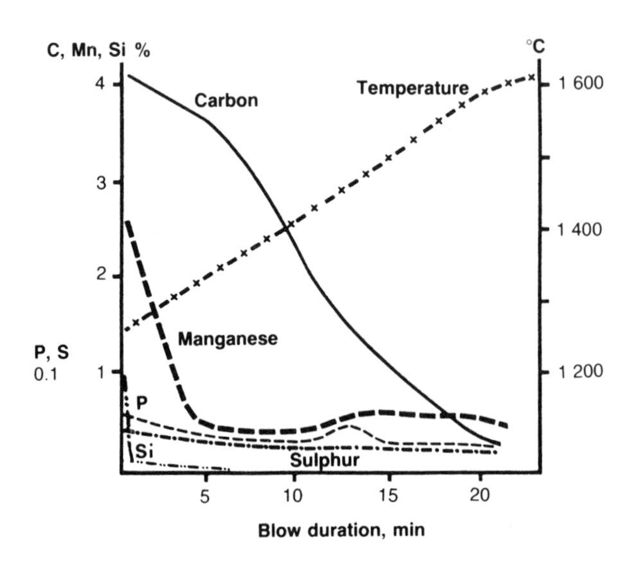

3.5 Removal of elements during a BOS blow

Carbon

Carbon is the most significant element in the effect it has on the properties of steel. Hot metal is virtually saturated with carbon on leaving the blast furnace but it may be reduced to one hundredth of saturation level during steelmaking using oxygen.

Since the advent of tonnage oxygen carbon removal has been straightforward in steelmaking and the rate of the reaction

$$[C] + [O] \rightarrow CO \qquad (18)$$

is related to the carbon level in the molten metal

$$\frac{-d[C]}{dT} = k\,[COg]\,[C] \qquad (19)$$

Oxygen in gaseous form is an integral ingredient in BOS. However, prior to BOS, when electric arc and openhearth were the chief processes, the attainment of very low carbon contents in steel (less than 0.08%C) was difficult and time consuming. This was when oxygen was provided as iron ore or millscale. With gaseous oxygen injection, carbon can be reduced to 0.01% without difficulty, rapidly and consistently. Decarburisation in the open hearth was as low as 0.4%/hour. In BOS, rates in the range 12-16%/hour are achieved.

The burden of carbon removal can be reduced by using some DRI (directly reduced iron) in the charge - particularly in electric arc furnaces - since DRI contains only 1-1.5%C compared with more than 4%C in hot metal. DRI is produced from ore at sub-liquidus temperatures and hence gangue and some unreduced oxide remain. These have a deleterious effect on production rate and energy needs.

In converters it is quite difficult to obtain a metal sample during blowing, so the progress of decarburisation is monitored by calculating the removal rate from gas analysis and flow rate data and relating the results obtained to a computer model. The computer calculates the volume of oxygen which needs to be blown and on completion the oxygen lance is withdrawn and the vessel is tilted to allow a sample to be taken for analysis and the temperature to be checked. A short final blow and adjustment to the slag composition can be made if required.

Slag Basicity

Silicon, sulphur, phosphorous and manganese partition between slag and metal during steelmaking and the equilibrium achieved are dependant on slag basicity. There is no satisfactory scientific definition of basicity and during this century almost every physical chemist with an interest in steelmaking has devised his own empirical basicity measurement.

Basic steelmaking slags are complex oxide mixtures and it is generally accepted that these range in nature from compounds which are strongly basic to those which are strongly acid. It is not practicable to measure the actual degree of acidity or basicity of the oxides and they may be arbitrarily classified as shown in Table 3.3.

Table 3.3 Classification of oxides in slags

Strong Base	Weak Base	Weak Acid	Strong Acid
CaO	FeO	Fe_2O_3	SiO_2
MgO	MnO	Al_2O_3*	P_2O_5

*Al_2O_3 is amphoteric, ie it may behave as a "weak acid" or a "weak" base depending on the circumstances.

Slag basicity has to be quantified before it can be related to slag/metal partition ratios and although there are many ways of doing this they fall into two main categories.

Basicity Ratio

Basicity ratios are evaluated by adding together the basic constituents of slag and dividing these by the sum of the acid constituents. One of the least complicated methods of describing basicity was introduced by L Blum in 1901, ie the lime/silica ratio.

Lime/silica ratio $= \frac{\% (CaO)}{\% (SiO_2}$

In 1947 Flood and Forland introduced the "V" ratio as a measure of steelmaking slag basicity.

"V" ratio $= \frac{\% (CaO)}{\% (SiO_2) + \% (P_2O_5)}$

Other workers have introduced basicity ratios which attempt to take into account the different degrees of acidity and basicity of slag constituents by using empirical constants.

eg basicity $= \frac{\%(CaO)-1.18\%(P_2O_5)}{\% (SiO_2)}$

or $= \frac{\%(CaO)}{\%(SiO_2)+0.634\%(P_2O_5)}$

Nett and Excess Basicity

The concept of measuring the "nett basicity" of slag is not new. Essentially the measurement is obtained by subtracting the sum of the acid oxides present from the sum of the basic oxides. In 1934 Hertz and his co-workers suggested that:

basicity = $\%CaO-\{1.86\%(SiO_2)-1.19\% (P_2O_5)\}$

In 1946 Grant and Chipman introduced the concept of an "excess basicity" parameter and this approach is extremely useful when considering sulphur removal in both ironmaking and steelmaking.

Excess basicity or

$EB = (MeO) -2 (SiO_2) - 4 (P_2O_5) -2 (Al_2O_3) - (Fe_2O_3)$

Where (MeO) equals the sum of the bases CaO, MnO and MgO which are considered as equivalent in neutralising the acid constituents of the slag. Considerations of "excess basicity" are quite complex and out of the range of this monograph but a lucid evaluation of the criteria involved is included in Bodsworth and Bell's textbook "The Physical Chemistry of Iron and Steel Manufacture".

In an $MeO-SiO_2$ slag system 2 MeO, SiO_2 would correspond to a nautral slag composition. If additional (MeO) is present the slag would be basic and have an excess basicity. If the (MeO) content were lower than that necessary to neutralize the acid constituents then the slag would have a "negative" excess basicity and so would be acid. Basic steelmaking slags always have a positive EB whereas ironmaking slags are often on the negative or acid side.

An understanding of slag chemistry is essential if ironmaking and steelmaking reactions are to be understood. Indeed the late Professor R Hay of Strathclyde University spent a lifetime studying slag and his oft repeated philosophy was that "if you get the slag composition right the correct metal composition will follow".

Silicon and Manganese

A small proportion of the siliceous gangue and more than half of the manganese oxide charged to the blast furnace is reduced during ironmaking to the levels indicated in Table 3.2. Both elements have a higher affinity for oxygen than iron and so are readily removed during the early stages of steelmaking. For the same reason silicon and manganese are used, together with aluminium to deoxidise steel prior to casting. If a flush slag practice is not operated both elements tend to revert from slag to metal as steelmaking proceeds and furnace temperatures increase.

Silicon Removal

$$[Si] + [O_2] \rightarrow (SiO_2) \qquad (20)$$

Figure 3.4 shows that silicon drops to a "trace" level of below 0.01% very quickly during the first few minutes of oxygen blowing. The SiO_2 produced is neutralised and stabilised by CaO in the slag. The reaction is strongly exothermic and the proportion of scrap which can be consumed during BOF steelmaking is directly related to the silicon content of the hot metal charged to the converter.

During the 1980s British Steel improved the productivity and fuel consumption in their blast furnaces and reduced the average silicon content of the hot metal market from about 1% in 1981 to just over 0.6% in 1988. The above change in practice reduced the potential heat content of the hot metal entering the converter and so the percentage of scrap consumed dropped from about 30% to 20% during this period.

This change in practice brought two major benefits:

(a) A 30% increase in vessel life which resulted from a decrease in slag bulk.

(b) An improvement in steel quality by lowering the tramp element level by reducing the proportion of scrap consumed.

Manganese Removal

Like silicon, manganese is readily oxidised out of steel early in the process. This is unfortunate since manganese has a beneficial effect on steel as it strengthens plain carbon steel without causing embrittlement and at the same time improves the weldability and low temperature impact properties.

$$[Mn] + [\tfrac{1}{2} O_2] = (MnO) \qquad (21)$$

During steelmaking, manganese attains an equilibrium level which although related to the hot metal manganese content tends to fall to about 0.20%. There is therefore little point in going to the expense of making high manganese hot metal in the blast furnace. High manganese hot metal is also undesirable since, by forming a stable sulphide

$$[Mn] + [FeS] \rightarrow [MnS] + Fe \qquad (22)$$

Manganese lowers the activity coefficient and sulphur in iron and so reduces the effectiveness of the ladle de-sulphurisation techniques discussed.

Sulphur and Phosphorus

A considerable amount of research has gone into the chemistry of sulphur and phosphorus in iron and steelmaking and a vast volume of literature has been generated and is available for detailed consultation elsewhere.

It is generally agreed that three parameters are of particular significance in determining the extent to which these elements can be successfully partitioned between slag and metal and these are summarised in Table 3.4.

Table 3.4

Parameters favouring sulphur and phosphorus removal

To effect removal of	DESIRED LEVEL		
	Slag Basicity	Temperature	% Oxygen in Metal
SULPHUR	High	High	Low
PHOSPHORUS	High	Low	High

It can be seen from the above table that the conditions which lead to sulphur and phosphorus eradication conflict to some extent. Removal of both elements requires a basic slag and this, of course, explains why neither element can be removed in a furnace with an acid refractory lining.

Most of the steels listed in the steelmakers recipe books, ie BS 970 specify that neither the sulphur nor the phosphorus content should exceed 0.050%. More stringent customer demands, particularly for high grade pipeline steels, frequently require steels to be supplied to maximum limits of 0.005% and even lower if possible. Although the above levels seem very low it should be remembered that they are gravimetric percentages - equivalent volume fraction levels would be about four times greater in each case. To increase machinability however, free cutting steels contain 0.2-0.3% sulphur and up to 0.07% phosphorus.

Sulphur and phosphorus are partitioned between slag and metal and the object is always to achieve a "partition ratio" which is as high as possible.

Sulphur Removal

The concept of excess basicity (EB) was introduced earlier. In considering sulphur partitioning between slag and metal it is best to write the reaction in the ionic form, ie

$$[S] + (O^{--}) = (S^{--}) + [O] \qquad (23)$$

and the equilibrium control K_S for the above section

$$K_S \quad \frac{(a_{S^{--}})\,[a_O]}{[a_S]\,(a_{O^{--}})} \qquad (24)$$

For sulphur removal equation (23) should be encouraged to proceed from left to right and this is favoured by having:

1. A high activity concentration of sulphur in the metal, ie high $[a_S]$ related to amount of S present and raised by elements like C and P which form compounds with Fe and lowered by presence of, eg Mn and Cu which form compounds with a stabilised sulphur in metal.

2. A high oxygen ion concentration in the slag, ie high $(a_{O^{--}})$, and so this is not directly measurable but is related to the excess basicity or EB of the slag where EB is calculated according to Bodsworth as

   $EB = n\,(CaO) + n\,(MgO) + n\,(MnO) - 2n\,(SiO_2) - 4n\,(P_2O_5) - 2n\,(Al_2O_3) - n\,(Fe_2O_3)$

   where n is the number of moles of oxide per 100g of slag. Each mole of the basic oxide supplies one oxide ion to the solution whilst two oxygen ions are acquired by a mole of SiO_2 and so on.

   The relationship between the sulphur partition ratio and the EB is shown in figure 3.6 and the effect of the ferrous oxide content of the slag is shown in figure 3.7.

3. A low activity concentration of sulphur ions in the slag, ie low $(a_{S^{--}})$.

4. A low activity concentration of oxygen in the metal, ie low $[a_O]$ related to (FeO) which is low in the blast

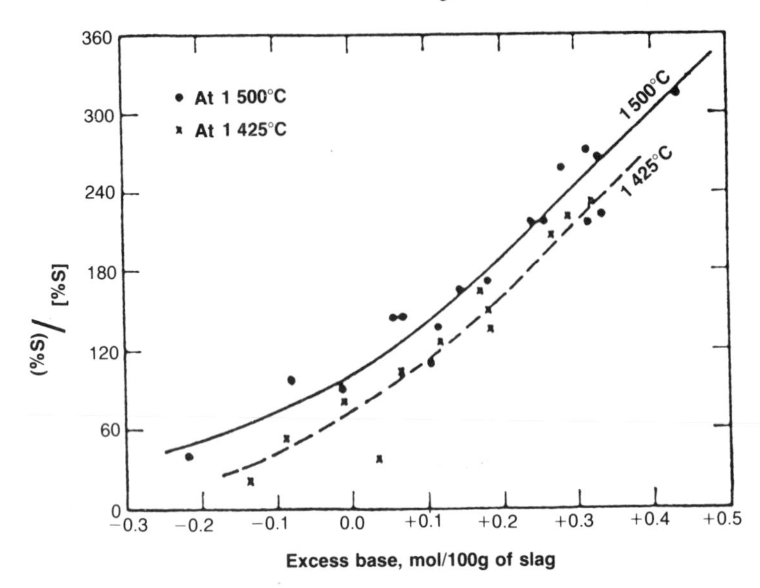

3.6 Relationship between "excess basicity" of slag and sulphur partition ratio (Hatch and Chipman)

furnace and during electric arc furnace "double slag" steelmaking which uses a second calcium carbide reducing slag to achieve high levels of furnace deoxidation.

Figure 3.7 shows the effect of (FeO) on the sulphur partition ratio at various levels of EB.

Phosphorus Removal

Phosphorus adversely affects the properties of most grades of steel and the tendency nowadays is to reduce the phosphorus content to levels of 0.010% and below. The conditions which favour phosphorus removal are high slag basicity, a high dissolved oxygen content in the metal and low temperature. In BOS steelmaking lime and oxygen are injected simultaneously and dephosphorisation occurs quite readily as shown in figure 3.5.

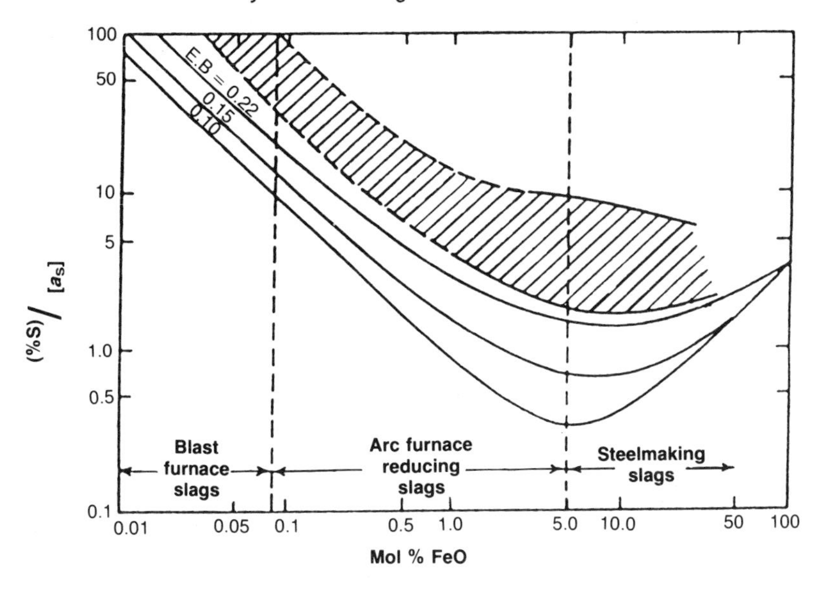

3.7 Relationship between sulphur partition ratio, excess basicity and ferrous oxide content of slag for various processes.

As steelmaking proceeds metal and slag temperatures progressively rise and some reversion of phosphorus from slag to metal may take place. Between 1980 and 1990 the proportion of steel continuously cast has tripled and one consequence of this is that the increased superheat requirement of steel made by this method has lead to a general increase in the tapping temperature of low carbon steel from about 1600°C to almost 1700°C.

Reversion of phosphorus can be prevented by removing the phosphorus rich slag early in the steelmaking process and while the metal temperature is relatively low. The LD-AC process was developed as a "flush multi-slag" steelmaking method which allowed low phosphorus steel to be made from hot metal with high phosphorus levels ranging from 0.4-2.0%. Alternatively phosphorus reversion can be reduced by forming a stable calcium phosphate compound in the slag but this might only be achieved by increasing a slag bulk and basicity to unduly high levels. Dephosphorisation is usually expressed in terms of the rection

$$2\,[P] + 5\,(FeO) + 3(CaO) \rightarrow (3CaO\;P_2O_5) + 5\,Fe \qquad (25)$$

but although the qualititave effects of the above reaction are widely accepted, quantitative evaluation of the variables give a wide scatter of results.

One reason for this is that slags in the ternary system P_2O_5-CaO-FeO have an extensive region of the liquid immiscibility so that in thermodynamic terms it is inappropriate to consider them as being ideal liquid.

DEOXIDATION

Steelmaking is an oxidising process and during BOS refining oxygen is injected into liquid steel at supersonic velocity and high pressure. At the end of refining liquid steel is saturated with oxygen which has to be removed before the alloy can be cast. If steel is not deoxidised before pouring the resultant alloy would contain blowholes of carbon monoxide and have extensive grain boundary network of iron oxide which would adversely affect the mechanical properties.

Figure 3.8 shows that the solubility of oxygen in liquid iron at the melting point of 1537°C is 0.16% whereas in solid iron at the same temperature the solubility falls to less than one fiftieth of that level, ie to less than 0.003%.

Because of the dramatic decrease in the solubility of oxygen in iron on solidification it is necessary to remove as much of the oxygen as possible while the steel is liquid. The solubility of oxygen is lowered by the presence of solute elements with an affinity for oxygen. Carbon reduces the solubility of oxygen in iron and further decreases in the level of oxygen in equilibrium with a particular carbon content are obtained by lowering the pressure. In practice steel is always chemically deoxidised to some extent by adding one or more solute elements which have a greater affinity for oxygen than iron.

In addition to chemical deoxidation special grades of steel are often subjected to vacuum and/or inert gas stirring treatments to minimise the inclusion content of the steel.

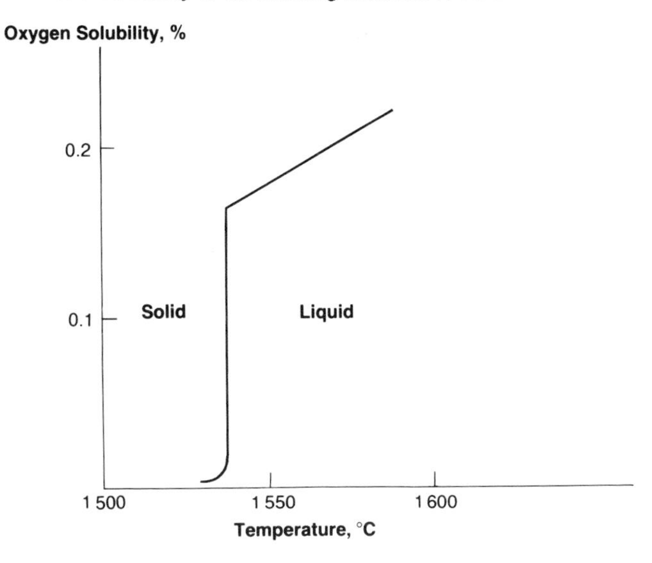

3.8 The solubility of O_2 in liquid and solid iron

Non Metallic Inclusions

Non metallic inclusions have a detrimental effect on the mechanical properties of steel in general but are particularly undesirable in alloys requiring good fatigue resistance. Steel contains both "indigenous" and "exogenous" inclusions. Indigenous inclusions are caused by the separation of non-metallic oxide and sulphide compounds as a result of solubility changes and other reactions which take place in the steel during cooling.

Exogenous inclusions are refractory particles which are picked up from furnace, ladle and tundish linings and the nozzles and holloware which contact the liquid metal.

Figure 3.9 shows that vacuum treatment lowers the dissolved oxygen content of iron in equilibrium with a given level of carbon so that vacuum degassed steel should have a lower indigenous inclusion count than non degassed steel.

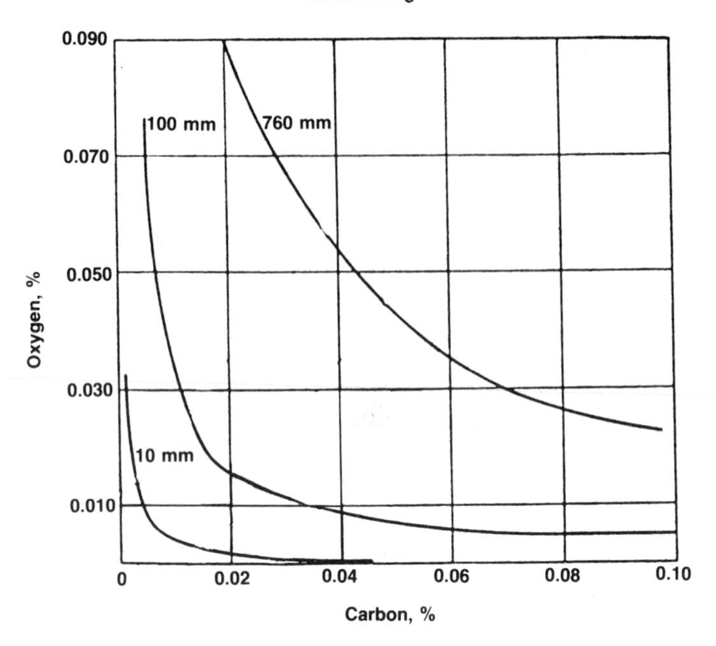

3.9 Carbon and oxygen equilibrium levels at various pressures

Unfortunately degassing involves tapping the furnace at higher levels of superheat which not only increases the dissolved oxygen content but can lead to greater refractory erosion and a consequent increase in exogenous inclusions. Although it is generally accepted that vacuum treatment leads to the production of cleaner steel, to some extent a swings and roundabouts situation is created in which the indigenous inclusion content might be reduced but the exogenous inclusion content might increase. Vacuum degassing does however change the nature and size distribution of the inclusions and the final process route is influenced by the properties required by the customer.

Obviously steelmakers strive to reduce inclusions to a minimum level and before this can be done it is important to be able to identify the source of the low metallic particles. This is not always as straightforward as it might appear since deoxidation and ladle erosion products are often aluminosilicate particles. Steel is deoxidised whilst liquid by adding elements which have a high affinity for oxygen and which are capable of forming oxide compounds before solidification commences.

Ideally the non-metallic oxides formed should have a low density compared to liquid iron and should float rapidly out of the steel once formed. After tapping liquid steel loses heat and becomes more viscous so as time passes the rate of flotation of oxide particles decreases. When the steel finally solidifies any non-metallic particles which have been trapped in the steel are termed "inclusions".

Deoxidation Practice

Conventional deoxidation may be regarded as the final stage in the steelmaking process. Electric arc furnace steel is often deoxidised prior to tapping whereas BOF steel is treated by adding deoxidants to the tapping stream and to the ladle. Some of the elements which are used to deoxidise steel are listed below.

Most common — manganese, silicon and aluminium

Occasionally used — titanium, zirconium, calcium, magnesium

Manganese, silicon and aluminium are the most commonly used deoxidants in steelmaking. All three elements are relatively cheap and benefit the properties of particular alloys of steel. Manganese and silicon have melting points of 1242°C and 1415°C respectively and to enhance their solubility in liquid steel they are generally added to steel as lower melting point ferroalloys. Aluminium melts at 660°C so the metal melts readily at steelmaking temperatures. Because of its low density and its tendency to float in liquid steel aluminium is often cast into annular "doughnut" or star shaped blocks and thrust under the surface of liquid steel on iron bars.

The effectiveness of individual deoxidants is related to the free energies of oxygen and the deoxidant dissolved in iron [ ], or slag ( ).

		ΔG @ 1600°C kJ/mol O_2
[Mn] + [O]	→ (MnO)	- 93.8
[Si] + 2[O]	→ (SiO_2)	- 163
2[Al] + 3 [O]	→ (Al_2O_3)	- 324

It can be seen from the above that silicon is a more effective deoxidant than manganese and aluminium is more effective than silicon. It has been observed that using more than one deoxidant has a beneficial effect on the activity concentration of other deoxidising elements.

Once formed it is vitally important that the products of deoxidation float out of the melt as quickly as possible. In the past many workers considered that the rate of flotation of inclusion was governed by Stokes's Law.

$$V = \frac{0.222\, g\, r^2\, \Delta e}{\eta} \text{ cm/sec} \qquad (26)$$

where
- V = rate of rise of inclusions cm/sec
- g = gravitational constant 981 dyne/gm
- Δe = difference in density of inclusion and slag g/cm²
- r = inclusion radius cm
- η - viscosity of slag dynes s/cm²

Consideration of the above equation suggests that for a high velocity of inclusion flotation and clean steel it is desirable to have relatively large inclusions of low density.

Until 1965 many steelmakers tried to control the composition of the inclusions formed in steel to ensure that they were liquid at steelmaking temperatures the idea being that such inclusions could coalesce and grow in size in the liquid steel and so float out of the metal at even higher speed. In 1965 Plöckinger and other workers advocated deoxidation with aluminium which produced small, solid, relatively dense alumina particles in steel. Plockinger doubted whether or not Stokes's Law considerations could be applied to the mechanism of inclusion flotation in steel and he claimed that although the inclusions formed by aluminium deoxidation were smaller and denser than steel deoxidised by manganese and silicon, the final product was at least as clean.

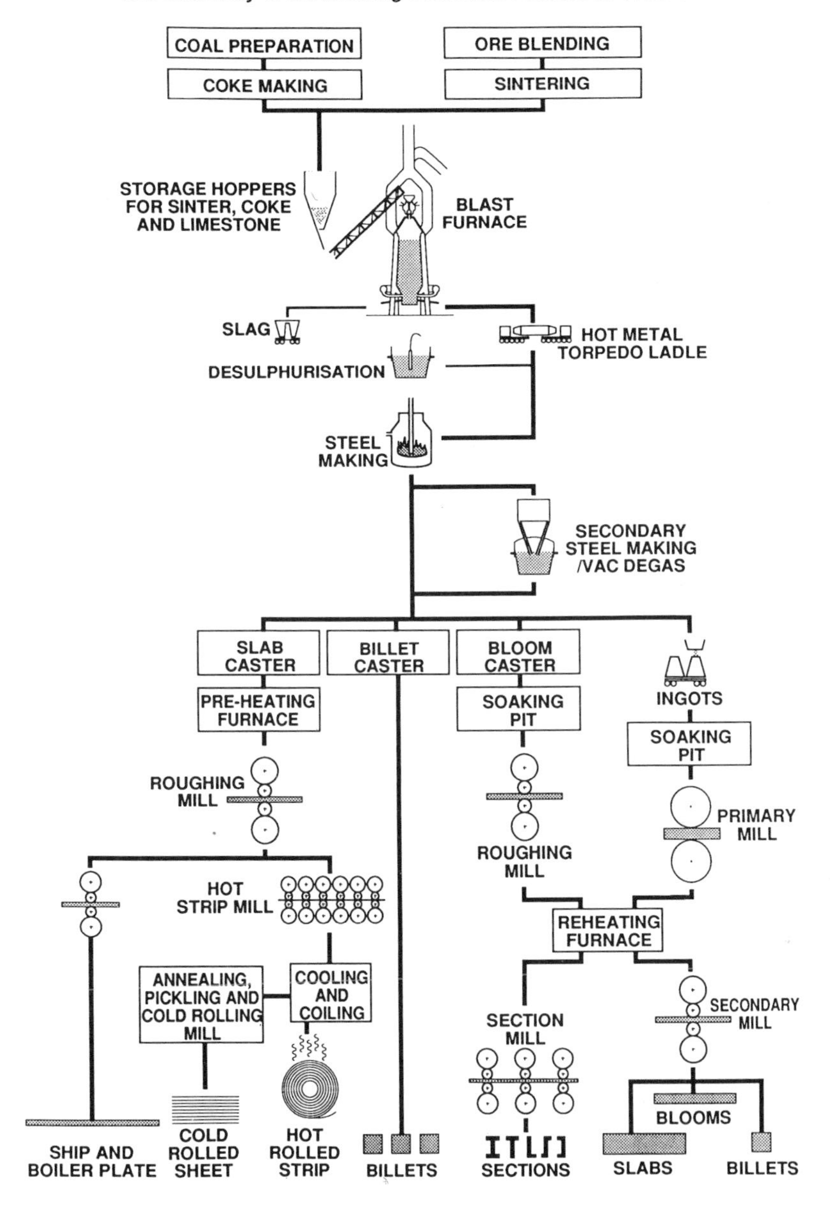

3.10 Schematic Process Flow diagram of a large integrated steelworks

There is no doubt that aluminium is capable of reducing the dissolved oxygen content of steel to a very low level and, following aluminium treatment, consistent and high recoveries of alloying elements are achieved allowing steel to be made to tight specification. Alumina inclusions can however cause problems by gathering and spoiling the stream of metal flowing from tundish nozzles particularly during continuous casting. In solid steel alumina inclusions are extremely hard and non deformable and are likely to cause breakages during wire drawing, tool tip blunting in drilling and machining, and are likely to spoil the inner surfaces of seamless tubes by their scoring and scratching action on the piecing mandrel.

Aluminium is also added to control the grain "size" of steel and leads to an improvement in mechanical properties. There is no "best" way of deoxidising steel and the method adopted must be appropriate both to the process route chosen and to the requirements of the customer.

Figure 3.10 shows a schematic diagram illustrating the flow of materials during the entire process route in a large modern integrated plant.

CHAPTER 4
Basic Oxygen Steelmaking
(BOS, OBM, Q-BOP, Combined Blowing)

INTRODUCTION

In 1988 world crude steel production was 778 million tonnes and in the West approximately 70% of this total was made by the basic oxygen steelmaking process. In the United Kingdom and elsewhere this method is usually referred to in abbreviated form as the BOS process or the BOP but in North America the term BOF (basic oxygen furnace) is more often used. In Europe the method is sometimes described as the LD or Linz-Donawitz process in recognition of the twin Austrian towns where early developments took place between 1947 and 1949.

In this monograph, the terms BOS, BOP, BOF and LD are considered to be entirely synonymous.

DEVELOPMENT OF THE PROCESS

The BOS process is a natural successor to the Bessemer Converter in that hot metal at about 1400°C is converted to steel at 1650°C by the exothermic oxidation of metalloids dissolved in the iron. No external heating is necessary and most of the heat generated in the process comes from the oxidation of silicon, manganese and phosphorus.

Tonnage oxygen was not available for steelmaking until after the Second World War so that the Bessemer and Thomas converters of the 19th century were air blown. The nitrogen in the air carries heat out of the system and lowers the thermal efficiency of the earlier processes. Whereas air blown converters cannot consume more than about 5% of cold steel scrap, which was added as a coolant to control tapping temperature, oxygen blown BOFs can use up to 30% of scrap. The preparation of scrap used in BOS steelmaking is obviously of economic importance and is related to the relative cost of hot metal and the availability and price of good quality, low residual scrap.

During the past 10 years the tendency has been for blast furnaces to be operated at lower silicon contents and this has resulted in the scrap consumed being reduced from about 30% to 20% for the reasons outlined in Chapter 3.

In the BOS process oxygen of greater than 99.9% purity is blown at supersonic velocity, through a water cooled copper, multihole lance onto the surface of the mixture of hot metal, scrap and fluxes in the steelmaking vessel. Oxidation of impurities takes place and at the end of oxygen blowing the composition of the charge is typically modified as follows:

Hot metal at 1400°C	4.6%C	0.7%Si	0.4%Mn	0.05%S	0.08%P
Steel at 1650°C	0.05%C	Trace Si	0.2%Mn	0.03%S	0.02%P

Figure 4.1 shows the sequence of operations during BOS and Figure 4.2 shows a typical vessel and its associated plant.

DEVELOPMENT OF A PRACTICE FOR TREATING HIGH PHOSPHORUS HOT METAL

The original LD process made steel from a charge which consisted of approximately 70% low phosphorus hot metal and 30% scrap steel. In the late 1950s lance design modifications allowed powdered lime to be injected with oxygen and this greatly speeded up slag formation and slag-metal reactions within the vessel. By using flush slag techniques, in which the initial phosphorus saturated slag is removed and a new slag is subsequently formed, the phosphorus content of the treatable hot metals was raised from a maximum of 0.4% in the straightforward LD furnace to 2.0% in the lime injection LD-AC. Typical analyses of the hot metal used in both processes is shown in the following table.

Basic Oxygen Furnace

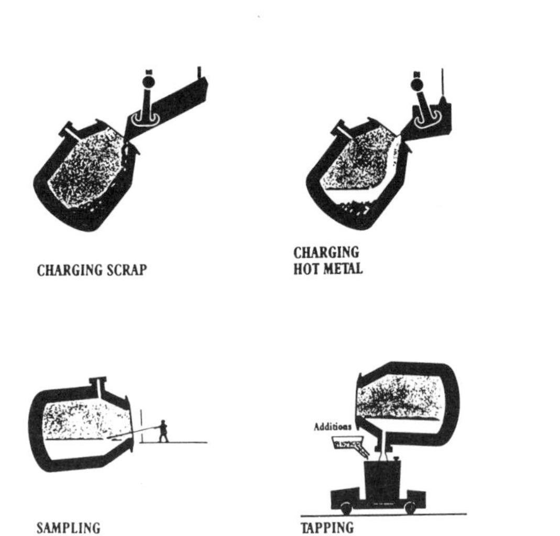

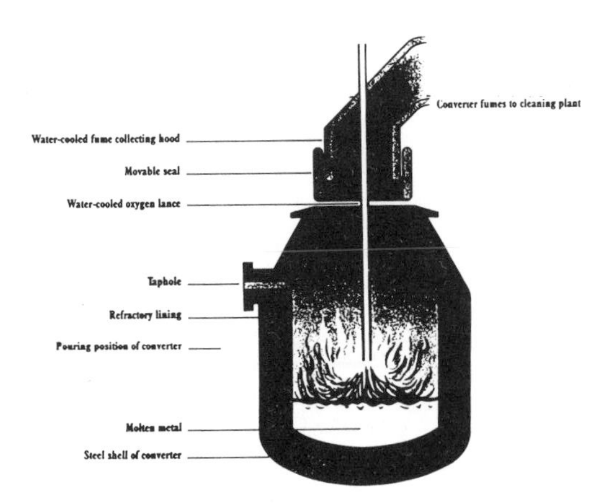

4.1 Sequence of Operations during basic oxygen steelmaking (Information Services, British Steel)

Table 4.1
Hot metal composition for LD and LD-AC practice

	Carbon %	Silicon %	Manganese %	Phosphorus %
LD	3.7 to 4.5	0.5 to 1.5	0.7 to 2.0	0.1 to 0.45
LD-AC	3.7 to 4.5	0.5 to 1.5	0.7 to 2.0	0.4 to 2.0

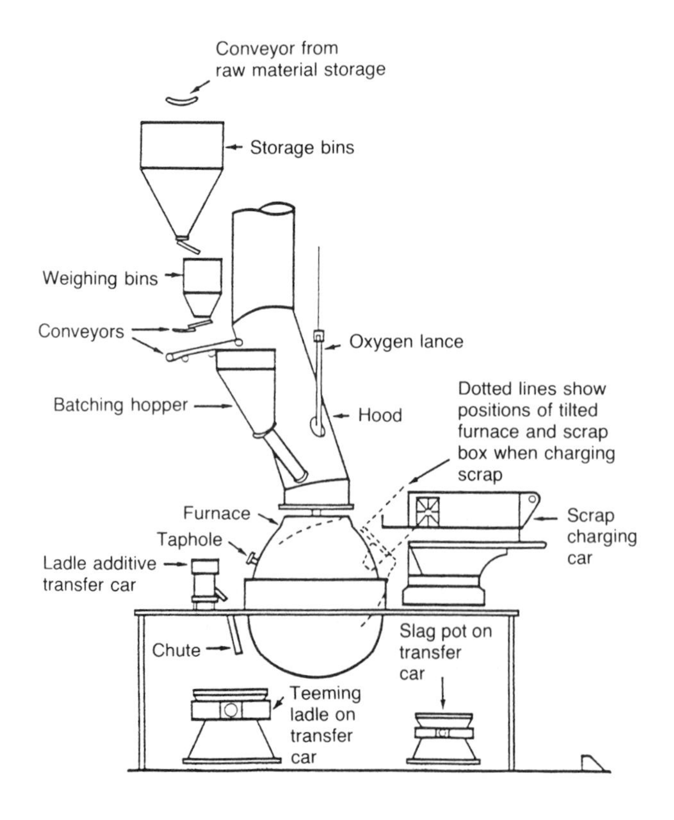

4.2. A BOS vessel and ancillary plant

The lime injection innovation was developed concurrently in Austria, Belgium, Luxembourg and France and its introduction was particularly beneficial in continental Europe where the indigenous ores give rise to high phosphorus hot metal.

Because of the increased demand for high grade weldable steels for stringent applications like offshore pipelines, acceptable levels for sulphur and phosphorus have dropped from 0.04% to less than 0.01% during recent times. The LD-AC process has consequently become less attractive as it is generally cheaper to charge the blast furnace with more expensive but lower phosphorus burdening materials.

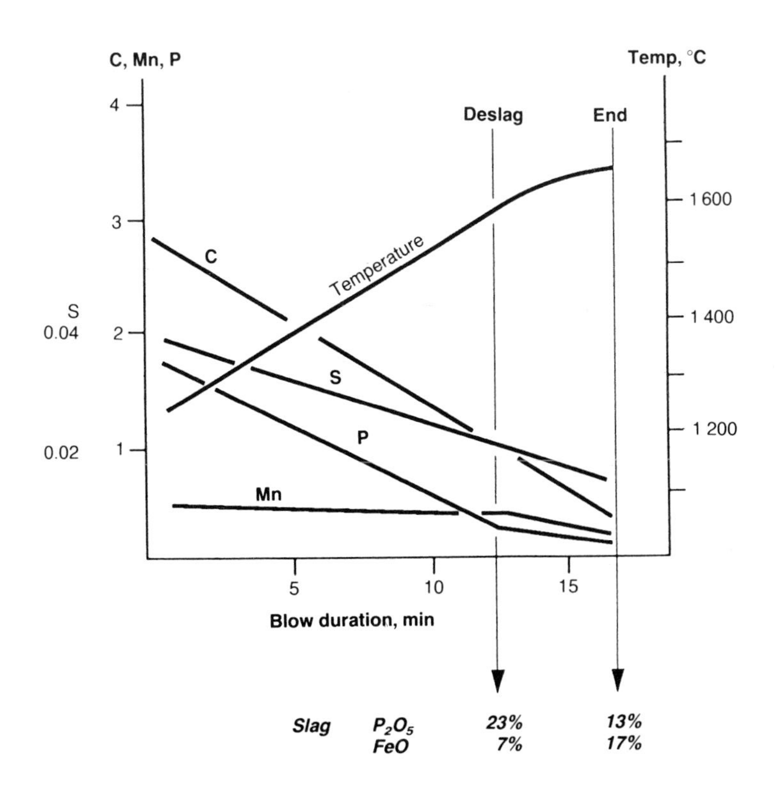

4.3 Refining curves for high phosphorus iron

Table 4.1 shows the heat and composition changes which take place during LD-AC steelmaking and figure 4.3 and figure 4.4 shows typical refining curves for 75% hot metal and 25% scrap.

Table 4.1

Composition changes in high phosphorus iron

LD-AC practice

Heat Details		Composition Changes			
		Element	Iron	1st Blow	End
Tapped weight t	200	C	3.6	0.8	0.05
Blow 1 min	25	Si	0.35	trace	trace
Blow 2 min	12	P	1.8	0.2	0.02
Scrap %	35	Mn	0.45	0.30	0.05
O_2 flow, Nm^3/min	325	S	0.06	0.04	0.02
O_2 used, Nm^3/t	60				
Lime, Blow 1 kg/t	80	Slag Fe		>7	25
Lime, Blow 2 kg/t	40	Slag P_2O_5		20	7
Lime (Total)	120				

Typical tap to tap time 60 min. Tapping temperature 1625°C. Production rate 200 tonnes/hour which is substantially lower than the output which would be achieved using a non flush slag straight BOS technique.

OPERATING BOS CONVERTERS

Steelmaking is an oxidising process and the reactions which occur in a BOS converter are the same as those which occur in any other basic steelmaking furnace although the dynamics of the reactions may vary because of the different conditions which prevail in alternative processes. After charging the furnace with hot metal, as shown in Fig 4.1, oxygen is injected

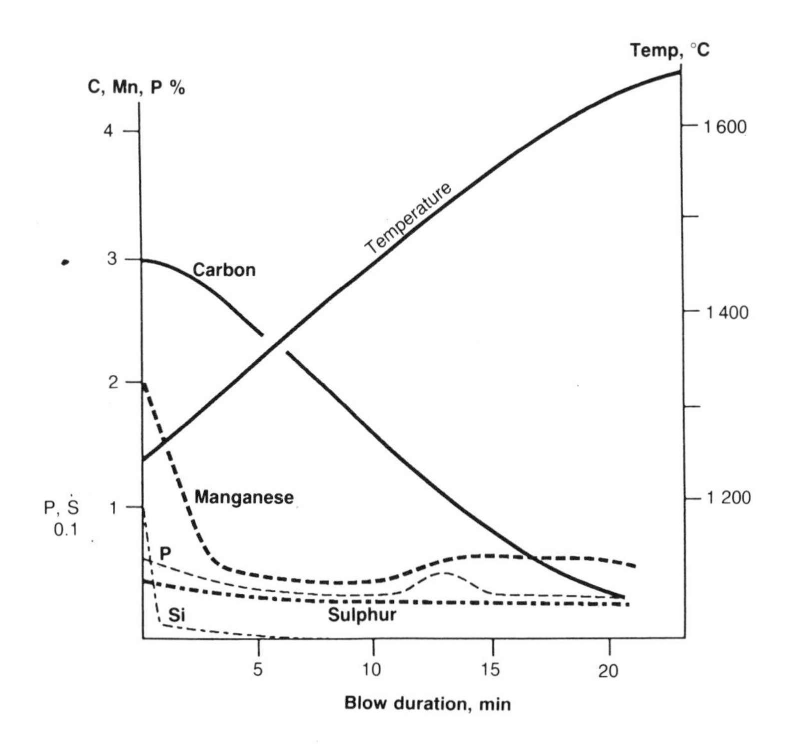

4.4 Refining curves for low phosphorus iron

into the vessel at a velocity of over Mach 2 using, on average about 55 Nm^3/t.

In a 300 tonne converter oxygen might be blown at carefully monitored rates for up to 20 minutes during which time the gas evolved is continuously analysed by infrared techniques cleaned and stored for reuse providing its calorific value is high enough. The actual blowing rate and lance height are monitored and controlled by a computor which although preprogrammed is linked by various controlling sensors as shown in figure 4.5.

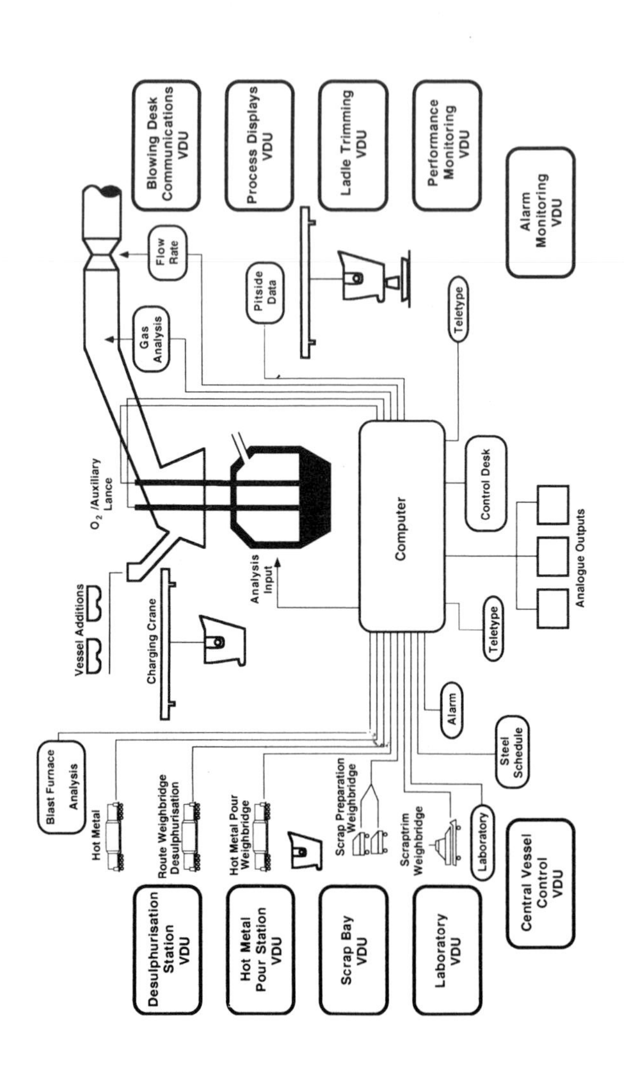

4.5 Steel plant computer Links

Lance height is regulated by a closed loop audiometric device which optimises and adjusts lance height as the blow progresses. It is necessary to control lance height to obtain the fast reactions which take place with good mixing of slag and metal. During the working of a melt the lance usually starts in the high position about 2-4m above the metal level and it is gradually lowered to 1m as the metal is refined. If the lance is "too high" the slag in the vessel foams and may overflow. Blowing with the lance in "too low" a position results in too much direct impingement of oxygen on the metal surface and as slag is pushed aside copious quantities of red, iron oxide fume may be generated and metal yield is decreased.

Injected oxygen contacts a proportion of the metal surface without extending the reaction area to the vessel walls which would result in rapid refractory wear. The metal in the contact area is rapidly oxidised and diffusion and absorption of oxygen and slag emulsification and mixing with the metal occur simultaneously. Decarburisation follows silicon removal and Fig 4.3 shows a plot of typical removal curves for carbon, silicon, sulphur, phosphorus and manganese together with the accompanying temperature change which occurs during the progression of a melt. The carbon monoxide generated during carbon removal ensures agitation and mixing of the metal and slag in the converter and carbon monoxide evolution remain brisk until the carbon content drops to about 0.1%.

The major problem with BOS steelmaking is in monitoring the temperature and composition of the metal during oxygen blowing. Over the years many methods have been tried and it seems that at last sub lances which can be lanced into the liquid metal to measure these parameters have become available. If furnaces have to be "turned down" to check temperature and metal composition, significant production losses ensue because of the time taken.

ENGINEERING OF BOF SYSTEMS

Steelmaking reactions occur in a dense, very hot, aggressive liquid which is in constant rapid movement and from which large volumes of very hot gases with a high dust content emerge. So that effective metallurgical control is possible, the engineering of the system has to contain the reacting system, deliver to it oxygen, scrap, hot metal, lime and ferroalloys and take from it gas, fume, slag and steel. To put the amounts to be handled in perspective, the raw materials and products from a vessel system million tonne/year (Mtpy) steel plant are given in Table 4.2.

Table 4.2

Consumables and products of 1/Mtpy steel plant

Materials Consumed

Liquid Steel	(95% yield)		1,050,000	tpy
Hot Metal	70%		740,000	tpy
Scrap	30%		310,000	tpy
Oxygen	56 Nm³/t		59 x 10⁶	Nm³
		or	80,000	tpy
Lime	115 kg/t		115,000	tpy
Refractories	3.5 kg/t		3500	tpy

****

Product

Steel			1,000,000	tpy
Slag	(15%)		160,000	tpy
Gas	65 m³/t		68 x 10⁶	Nm³
		or	91,000	tpy
Dust	1%		10,500	tpy

The whole of these quantities will pass through one of the two vessels whose nominal capacity for the example in Table 4.2 would be 100 tonnes.

The Vessel

A typical modern top blown BOF vessel is shown in figure 4.6. The modern vessel has two other significant differences from early types. These are:

1. A tap hole is used to avoid lip pouring. This reduces refractory wear at the mouth and aids the clean separation of metal and slag during tapping.

2. Specific vessel volume has increased so that modern vessels have a volume of about 1 m³/tonne of steel.

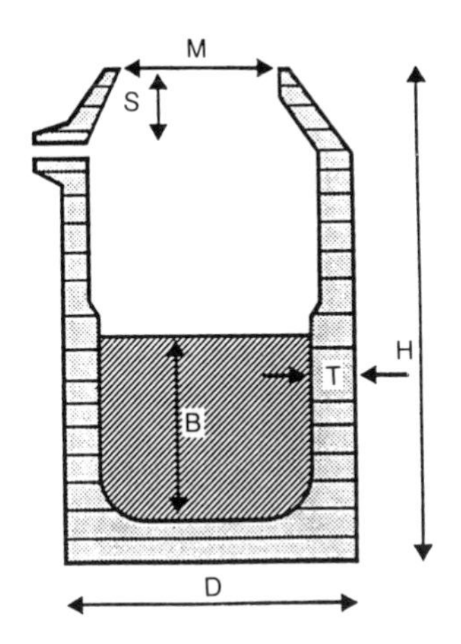

Parameter	Symbol	Ratio	Typical value
Capacity	C		250t
Volume (working)	W	W/C	$1m^3/t$
Volume (shell)	V	V/C	0.9
Height	H		12m
Diam.	D	H/D	1.5
Thickness	T	H/T	10
Bath depth	B	H/B	6
Mouth diam.	M	H/M	3
Tap hole depth	S	H/S	3.5

4.6 The modern BOF vessel

At the present time the largest vessels in operation are about 400 tonnes. The practical limit on size is due mainly to the problem of tilting vessels holding 400 tonnes of liquid steel since when lined and full the tilting mass may well approach 1,500 tonnes. Bath depths of more than 2.5m provide difficult reaction conditions so that the larger the capacity the squatter the profile. This leads to a more difficult stress pattern to be dealt with in the vessel. The plates forming the shell can reach temperatures of 200-300°C so that careful design is necessary to avoid problems caused by vessel distortion during tilting. Highly rated and powerful electric motors of several hundred horsepower are needed to tilt the vessels in the short times required for high production rates.

A very important aspect of the vessel is the lining which contains the steel. High grade dolomite linings are used and lining life is influenced by production rate, hot metal practice, steel quality, slag control and gunning or repairing practice. Linings wear by erosion, dissolution and abrasion but with good practice lives of over 1,000 heats are possible.

Masses in the range 4-7 kg of refractory material are generally consumed per tonne of steel produced but under favourable conditions consumption may be reduced to 2 kg/tonne.

The major changes that have occurred during the development of BOF steelmaking are given below in Table 4.3.

Table 4.3

Development in BOF output rates

Vessel size tonnes	35t to 450t
Blowing rate	2.5 to 4.5 Nm^3/t/min
Blowing time	22 min to 13 min
Tap to tap	60 min to 30 min
Availability	50% to 75%
Output	55 to 600 t/h

Gas, Dust and Heat

During oxygen blowing the above products are emitted from the vessel at a high rate. Off blow, the rate of emission is low so that the vessel extraction system has to be designed to cope with low, high and variable rates of emission.

Oxygen blowing rates up to 60,000 Nm^3/h require systems capable of dealing with 180,000 m^3/h of gas. Waste heat boilers have been used to recover sensible heat from the gas but are now rarely installed.

With current suppressed combustion systems and fume cleaning plant, a high proportion of the dust and calorific value of the gas can be recovered for reuse and at the same time contribute significantly to a cleaner environment. The total energy in waste gas is approx IGJ/t and 70% of this is recoverable as unburned CO. The dust burden is 15-20 kg/tonne, most of which can be separated and used as a coolant in BOS or as a source of iron for the blast furnace via the sinter plant.

Gas collection is controlled by CO content. Typically gas with less than 25% CO is cleaned and then flared. More than 25% CO gas is cleaned, cooled and stored in a holder with a capacity for about 3 heats. The gas is then compressed into the gas distribution system. 75 to 100 Nm^3 of gas/tonne of steel can be recovered with a calorific value of about 8.5 MJ/Nm^3. With the current high cost of oil and natural gas, the use of BOS gas is very cost effective.

Dry dust is collected in two stages. The primary stage collects about 40% of the total as coarse dust which contains 75-80% iron. This can be used as a coolant in the BOS furnace. The balance of fine dust is collected from electrostatic precipitators with only about 20% of metallic iron. This dust may be briquetted and returned to the sinter plant. A modern dust and gas recovery system is shown in figure 4.7.

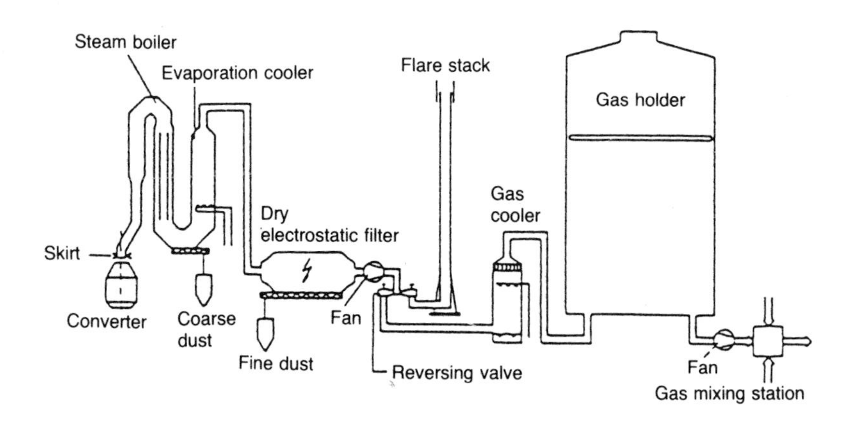

4.7 Converter gas and dust recovery

Wet dust systems require sludge treatment and disposal and recycled water treatment and will in the long term be environmentally less acceptable and probably higher in operating cost than dry systems.

ESSENTIAL FEATURES OF BOF SYSTEMS

The BOS converter has given good service since its acceptance in the 1950s. Techniques have improved during the years and with prerefining, vacuum degassing, secondary steelmaking and continuous casting excellent quality plain carbon and low alloy steel can be made by the process. Up to 80% of the charge to the furnace is hot metal from the blast furnace so that it seems that BOF steelmaking will remain viable so long as good quality coking coal is available at a reasonable price for ironmaking.

1. A blast furnace as a source of hot metal is essential since up to 80% of the BOF charge is hot metal.

2. Because a blast furnace is needed and that furnace uses large quantities of high quality coke, problems will intensify as supplies of such coke diminish. The result will be a higher cost product from the BOF.

3. Very high production rates (up to 600 t/h) can be reached using large vessels. Those levels of production, however, can only be achieved by making several hundred tonnes of one specification and markets must exist for cast quantities.

4. Individual vessel production is high and tap to tap time can be as low as 30 minutes, provided that the metallurgical load on the BOF has been reduced prior to BOF processing.

5. The rapid reactions and the arduous environment make process control difficult. It is in this area that most improvements have to be made.

6. The process produces a considerable amount of energy in the form of hot inert gas and burnable carbon monoxide. Waste heat boilers to recover the former are rarely installed, but combustible gas recovery has become standard.

7. The process is particularly suitable for the high tonnage low carbon steels for strip but other carbon and microalloyed steels are processed.

THE OBM, Q-BOP PROCESS

Bessemers ideas of the 1850s were for a bottom blown converter that used air to decarburise iron and make steel. His original patents included the use of oxygen but its use to enrich air for bottom blowing caused a significant decrease in bottom refractory life. Since some air was still used, the higher temperature generated by added oxygen, caused pickup of nitrogen making the process unacceptable metallurgically for a wide range of steels.

In the late 1960s, the problem of injecting pure oxygen into the base of steelmaking vessels was overcome. This was done by surrounding the oxygen stream with an annular jet of a hydrocarbon. The hydrocarbon cracks endothermically and keeps the bottom refractories at an acceptable temperature which makes the life of the converter base economically viable. By 1970, some 5 million tonnes of steel were made by the route and by 1980, 60 million tonnes. With the downturn in the world steel industry since that date, current bottom blown steel capacity will be of the same order as in 1980.

Hydrocarbon Shielding of Oxygen

As has been already mentioned early attempts to increase the production rate of acid and basic Bessemer furnaces, by oxygen enrichment of the air blast, failed because of accelerated refractory wear in the tuyere zone, although some measure of success was obtained in converters using oxygen in conjunction with coolants such as steam and CO_2. In the mid 1960s L'Air Liquide of Montreal, Canada developed a new concentric tuyere which allowed oxygen surrounded by an annulus of inert gas, to be blown through a tuyere as shown in Fig 4.8.

In the LWS process where fuel oil is used as an alternative to gas, additional heat is absorbed in vaporising the oil prior to cracking. The heat absorbed in cracking various hydrocarbon gases is given in Table 4.4 but oxygen and slagmaking additions may be injected through 20 or more tuyeres set in the furnace bottom.

Table 4.4

Cracking energy required for various hydrocarbon Gases

Hydrocarbon Consumed	Cracking Reaction	Heat
Methane	$CH_4 \rightarrow C + 2H_2$	75 kJ/g mol
Propane	$C_3H_8 \rightarrow 3C + 4H_2$	104 kJ/g mol
Butane	$C_4H_{10} \rightarrow 4C + 5H_2$	126 kJ/g mol

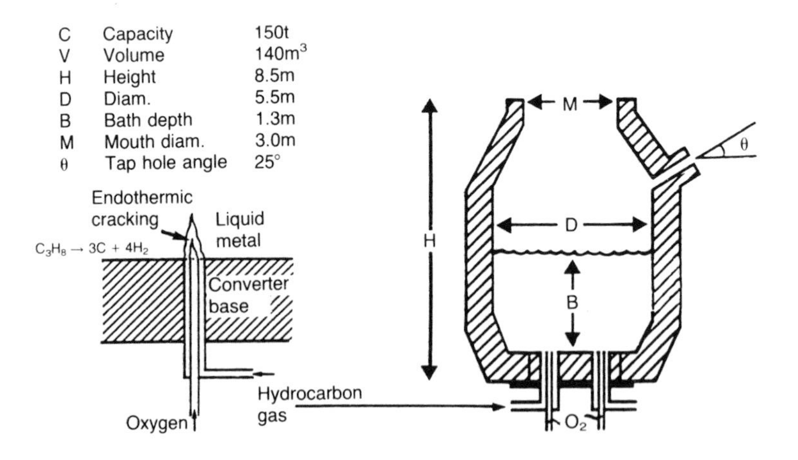

4.8 Typical bottom blown converter and concentric tuyere

Historically the processes were developed as follows:

The OBM Process

Dr K Brotzmann and company workers from Eisenwerke-Gesellschaft Maximilianshutte in association with Savard and Lee of L'Air Liquide, Canada successfully transformed the oxygen-bottom blown-Maxhutte (OBM) process from the experimental to full production stage between December 1967 and March 1968. In the OBM Process the oxygen entering the converter bottom is shielded with propane or natural gas.

The Q-BOP Process

Following a licensing agreement reached in December, 1971 between Maximilianshutte and the United States Steel Corporation, the North American equivalent of the OBM process was trade named the "Q"-Basic Oxygen Process "Q-BOP" by the USS Chairman, Edwin H Gott. The significance of the "Q" in the name is not known but it has been suggested that it indicates a quiet, quick and quality steelmaking process.

Q-BOP and OBM are very similar processes and a typical 150 tonne vessel is shown in figure 4.9.

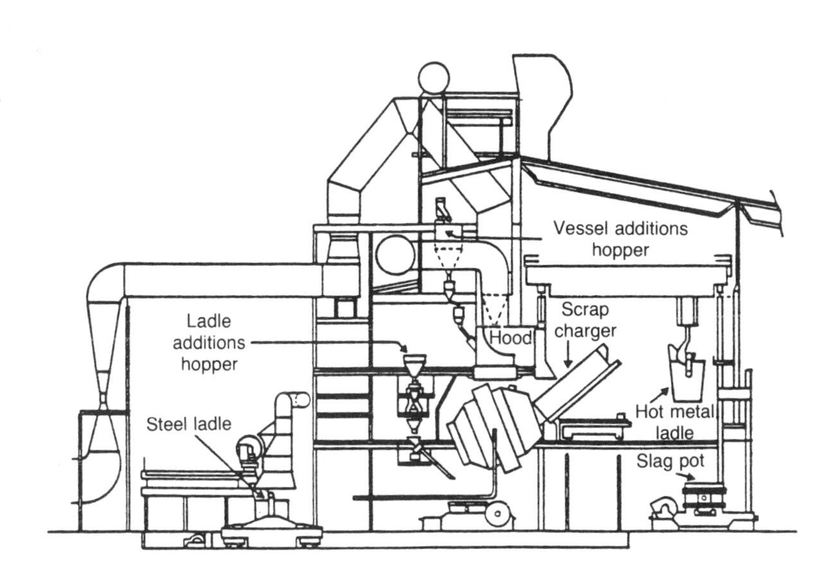

4.9 Bottom blown converter installation

In 1974 200 tonne Q-BOP units commenced production, in the United States, at Fairfield and Gary as open hearth furnace replacements whereas relatively small, less than 50 tonne OBM vessels, were installed in Sulzbach-Rosenberg and Volkingen between 1968 and 1974 to replace obsolete Thomas converter plant.

METALLURGY OF THE QBM, Q-BOP PROCESS

As in the BOF steelmaking process greater than two-thirds of the charge to the converter is hot metal from the blast furnace so that steelmaking consists essentially of the oxidising out of excess carbon, silicon and phosphorus from the charge and eliminating sulphur by control of slag basicity and the degree of metal oxidation. Since the metallurgy of steelmaking has been dealt with in Chapter 3, this section relates to essential differences peculiar to bottom blown processes.

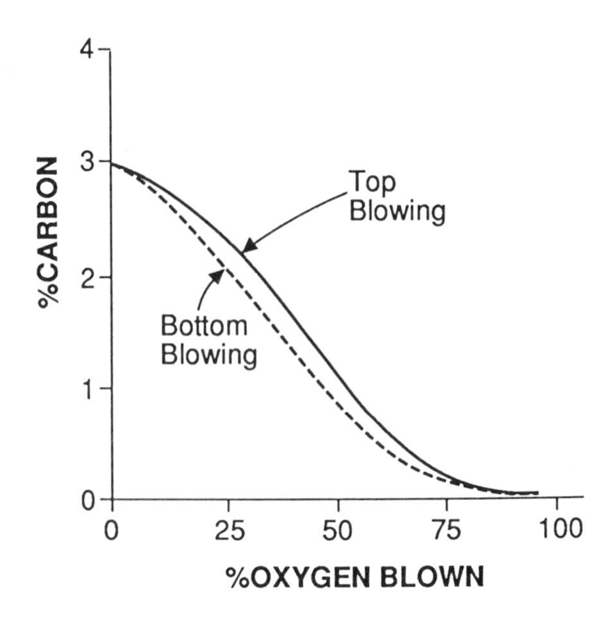

4.10 Removal of carbon during top and bottom blowing

Carbon

As in all pneumatic steelmaking processes carbon removal presents no problems and figure 4.10 shows that the rate of removal is very similar to that obtained in a top blown BOF converter. It has been claimed that very low carbon levels varying from 0.02 to 0.04% are easier to obtain in bottom blowing and that all carbon steel grades varying from 0.02% to 1.0% carbon can readily be made by the process. Once oxidised to carbon monoxide little combustion takes place within the steelmaking vessel consequently sensible and chemical heat is lost from the vessel.

Phosphorus

The highly phosphoric European irons often require the operation of a flush slag process. As in the LD-AC process a highly phosphoric slag is removed after an initial blowing period and the final slag for sulphur control, which is rich in iron and lime, is retained in the furnace after tapping and recycled. In the Q-BOP furnaces operated in the United States steel has been successfully made from low phosphorus hot metal.

Recent lance developments enable easily assimilated powdered lime and other fluxes such as fluorspar to be injected through the tuyeres together with oxygen.

Nitrogen

Nitrogen is especially undesirable in low carbon strip steel used for deep drawing applications. In bottom blown converter steel air is not drawn into the vessel to any great extent so that atmospheric nitrogen pickup is minimal.

Nitrogen can, however, get into the steel via:

1. impure oxygen - as in the BOS process the final nitrogen content of the steel is related to the purity of the injected oxygen.

2. nitrogen in the shrouding gas - Dutch natural gas may contain 14% nitrogen which can raise the final nitrogen content of the steel from 0.0018% to 0.0033%.

If strip steel for extra deep drawing applications is required the nitrogen content of the steel is minimised by using high purity oxygen and shrouding gas. If structural steel sections or railway materials are being produced however, high nitrogen contents are acceptable and the nitrogen may be considered as a hardening alloying element.

Hydrogen

Hydrogen is undesirable in low alloy steels, particularly forgings of substantial cross section, because of its effect of promoting hairline crack formation.

The unique feature of the processes under consideration is that they employ hydrocarbon shrouding and cooling media and unless inert gas flushing is employed at the end of a blow, residual hydrogen levels may be high. Experimental work on production plants has shown that the hydrogen level of steel made from both high and low phosphorus hot metal can be consistently reduced to a safe level of below 3ml/100g by argon or nitrogen flushing at the end of the blow. Nitrogen flushing might, of course, give rise to a higher nitrogen content in the steel and should not therefore be used for low carbon strip steels in which hydrogen would not be a problem in any case.

ESSENTIAL FEATURES OF THE OBM, Q-BOP PROCESS

The essential features of the processes can be summarised as follows:

1. Quieter blowing in OBM type vessels allows the utilisation of smaller capacity bodies than in comparative capacity BOS units. Less freeboard is necessary and it has been claimed that lower slopping losses result in yield increases of up to 2.5%.

   Further small gains in yield are obtained in bottom blown converters by virtue of the lower FeO content of furnace slag.

2. Conflicting data exists on scrap utilisation in OBM type vessels. BOS converters are able to consume about 30% scrap and it seems that converters of the former type

are able to melt between 25 and 36% during straightforward operation.

Scrap consumption could be increased by the employment of oxy-fuel burners.

3. Higher oxygen injection rates (3 compared to 2.5 Nm^3/min/tonne) in OBM compared to BOS converters result in faster steel production rates. For example, at the Fairfield Works of the United States Steel Corporation tap to tap times of 35 min have been obtained compared to 40 min for BOS furnaces of the same size.

   Oxygen utilisation is also better in bottom blown converters and savings of 6% of total oxygen blown have been claimed.

4. Powdered lime and fluxes can be injected into the vessel together with oxygen resulting in rapid slag formation and rapid slag/metal reactions.

5. Bottom blown converters produce less, but finer, fume than top blown converters.

   Small fume particles are difficult to remove from waste gases so that dust collection is probably equally expensive in both types of production.

6. Bottom blown converters are effective in obtaining very low levels of carbon so that steel with as little as 0.015% C can easily be made.

7. Refractory wear in both types of converter is similar with lining lives of 1500 heats/lining and better being obtained depending on practice.

8. Bottom blown converters are particularly useful for building into existing melting shops and below existing crane tracks.

   The OBM process has been installed in Europe to replace Thomas converters and the Q-BOP in the United States to replace obsolete open hearth furnaces.

COMBINED BLOWING

The main effect of the different methods of oxygen injection inherent in top and bottom blowing is the stirring intensity in the bath. It is much better in bottom blown converters and the slag/metal reactions approach equilibrium. There is little foaming of slag without lime injection, and with it the converter volume is of the order of $0.6m^3$/tonne. The result of top blowing is a very intense zone of reaction at the lance tip, with lesser rates of reaction near the vessel walls. Significant foaming of slag and metal occurs requiring vessel volumes in the range 0.7 to $1m^3$/tonne.

A secondary effect is that CO produced in top blown vessels is combusted in the vessel and that heat is returned to the charge. For bottom blowing, although there is less CO released from the bath which is near to equilibrium, that CO is combusted to a lesser degree than for top blowing and energy is lost. The calorific value of gas produced in both cases is 2000 kcal/m^3. Gas volumes produced are typically $80m^3$/t for BOF and $120m^3$/t for Q-BOP.

There are therefore advantages to be gained by bottom stirring of BOF to promote nearer equilibrium conditions and reduce slopping. There are advantages in top blowing some of the oxygen in a Q-BOP to combust the CO and obtain a better energy balance or in using more scrap in the charge. Since less oxygen is blown through the base, the life of the convertor base is improved. In 1985, 18 units of Q-BOP or LD were reported to be using combined blowing in production.

Combined Blowing in a BOF

The major part or all of the oxygen is blown via the top water cooled lance. Production of steel is now being done with bottom injection of:

1. Ar and N_2 through porous plugs or metal clad bricks

2. CO_2, Ar, N_2

3. 2-10% of O_2 with hydrocarbon shrouding

Gas injection rates of 0.1-0.2m³/min/t are used and the gas composition can be preprogrammed and controlled to suit the steel being made. Several years of operation have confirmed the following benefits from combined blowing in BOFs ranging in capacity from 100 to 330 tonnes.

1. Yield increased 1%

2. Ferroalloy savings up to 15% due to lower controlled state of oxidation of the bath.

3. Increase in lining life from the 500 to 1100 range up to 830 to 1400 heats.

4. Higher and more consistent control of composition with particular reference to carbon, sulphur, phosphorus and hydrogen.

Combined Blowing in Q-BOP

In Q-BOP, the addition of a top oxygen lance (and sometimes a side lance) is used to blow oxygen in a more gentle manner than in the BOF to combust CO to CO_2 and return that energy to the bath. Trials and production using 10-70% of the oxygen needed via a "soft blow" top lance have been done. The major advantage of this technique is to increase the scrap melting capacity of 50-70 kg/t over the standard Q-BOP process. Once more than 50% of the O_2 is being blown from the top, is the process still a bottom blown process?

THE FUTURE OF OXYGEN STEELMAKING

The evolution of the BOF from the pure top blown vessels of the 1950s and 1960s and the birth of pure bottom blown Q-BOP in the 1970s have now merged into the hybrid of combined blowing. This will continue but be dominated - as production is now - by the bottom stirred BOF rather than the top blown Q-BOP. The combined blowing hybrid has advantages over both pure top or pure bottom blowing all of which add up to a versatile, very cost effective method of steelmaking.

High rates of scrap usage are an added advantage and the use of cheap fuels to increase scrap utilisation to 40 - 60% of charge are being used by BOF and Q-BOP operators. In both cases the injection of granulated or powdered coal with more oxygen will increase the scrap melting capabilities of the vessels. Using the bottom tuyeres of Q-BOP as burners 100% scrap charges have been melted.

Perhaps the steelmaking unit of the future is a closed, pressurised vessel using hot metal and a higher portion of scrap than current BOF and perhaps up to 50%. Oxygen will be mainly top blown with argon, nitrogen, oxygen injected at the bottom. Coal will be added with one of the oxygen streams. High calorific value gases with minimum dust burden will be collected at pressure.

CHAPTER 5
Electric Arc Furnace Processes

Small electric arc furnaces were in use in about 1900 to manufacture tool steels and these small door charged Heroult designed units were installed to replace crucible steelmaking. By the 1920s top charging through swing aside roofs had been introduced on furnaces of up to 30 tonne capacity. These furnaces were powered with 7.5 MVA transformers to produce alloy steels in competition with open hearth furnaces. By the 1950s 90 tonne furnaces with 20 MVA transformers were making mild and plain carbon steels and the electric arc furnace became a modern steelmaking unit able to produce a wide range of steel grades. Such a unit had tap to tap times of about 4.5 hours and a corresponding production rate of about 20 tonnes/hour.

Currently, with ultra high power (UHP), oxy-fuel assisted melting, and oxygen lancing, 140 tonne furnaces melting mild and low alloy steel have tap to tap times of about 1.5 hours and production rates approaching 100 tonnes/hour. Straightforward arc furnaces are seldom used nowadays to make bulk stainless steels as these can be more efficiencly made by associating arc furnace melting with secondary refining as discussed in Chapter 6.

This chapter will, therefore, be concerned with the arc furnace as a high production unit for quality carbon and low alloy steels. The arc furnace is shown diagrammatically in figure 5.1.

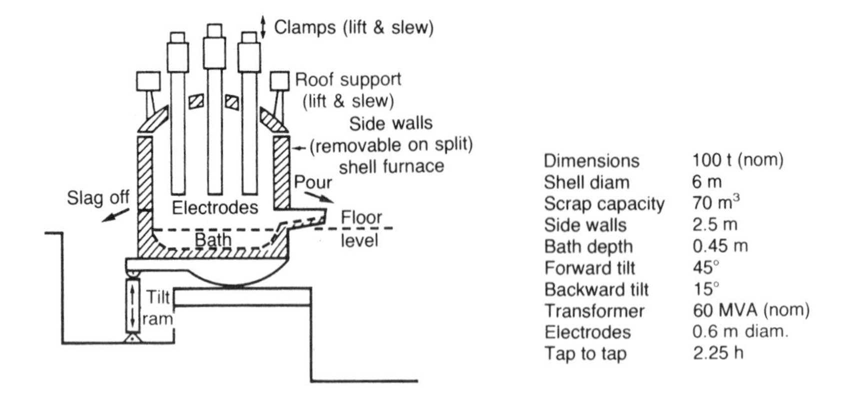

5.1 The arc furnace

THE METALLURGY OF ARC FURNACE STEEL PRODUCTION

Precise temperature control and clean melting in a sulphur free atmosphere made the electric arc furnace an attractive proposition for the melting of quality alloy steels. Arc furnaces are charged with clean scrap, steel and pig iron, limestone and possibly anthracite or broken electrodes as a source of carbon and melted as quickly as possible. Nickel and molybdenum, when required, are added with the scrap as these elements are not oxidised out during refining. At melt out, the bath is sampled for analysis and temperature determined. Bath temperature would be raised to 1530°C-1550°C whilst the analysis is being obtained and studied to decide on the metallurgical actions required to complete the heat.

Since there is an inherent, random variation in composition of the scrap, there is a similar variation at melt-out. With good practice the melt-out carbon should be at least 0.2 to 0.3% above the specification; sulphur and phosphorus are generally low and below 0.03%. If alloy steels are being made scrap is selected to ensure that the bath contains some of the alloys required, eg nickel, molybdenum and chromium. Such practice reduces the amounts of expensive ferroalloys needed to bring the steel into specification. Silicon and manganese are always present in steel scrap and plate iron and the quantity at melt-out will depend on the nature of the scrap and the relative amounts of plate iron and scrap charges. Plate iron is usually 75mm thick and can be contaminated by blast furnace slag and the plating bed sand.

Oxygen lancing follows melting.

Refining

Carbon is the principal element removed by the oxygen blow but other elements in minor quantities are also removed. These are silicon, manganese, phosphorus and chromium and each can be treated as an equivalent amount of carbon. The carbon equivalent for these elements is given below in Table 5.1.

Table 5.1
Carbon equivalents for oxidation

Element	Carbon equivalent % of 1% of element
Silicon	0.85
Manganese	0.22
Phosphorus	0.97
Chromium	0.35

Oxidation occurs in the manner already discussed in previous chapters and once again carbon removal and carbon monoxide evolution produces the "carbon boil" which is an essential feature of all steelmaking processes. The boil promotes stirring which results in good slag/metal mixing, elimination of temperature and concentration gradients, and the reduction of dissolved hydrogen and some nitrogen in the melt.

For these reactions, it is possible to calculate the theoretical amount of oxygen needed to remove 0.01% of each element and the resulting temperature rise. The data so calculated is given below in Table 5.2.

Table 5.2

Theoretical amount of oxygen needed and temperature rise resulting from oxidation of 0.01% of elements

Element	O_2 needed m^3/0.01%/t	Temperature rise °C/0.01% element oxidised
Iron	0.021	0.5
Carbon	0.099	1.3
Silicon	0.083	3.1
Manganese	0.021	0.8
Phosphorus	0.096	2.0
Chromium	0.035	1.2
Sulphur	0.075	0.8

This data is useful in showing relative amounts of oxygen needed together with the corresponding temperature rise resulting. In practice, the amount of oxygen needed depends on the initial carbon content and temperature of the bath. The temperature rise is dependent mainly on furnace size and rate of oxygen input.

Oxygen injection also increases the iron oxide content of the slag. Whilst this is not significant above about 0.2%C, below this level of carbon in the bath, the oxygen requirement for decarburising increases significantly. This is shown in figure 5.2 where the specific oxygen consumption in m^3/0.01%/tonne is given for bath carbons between 0.05% and 0.6%. Above 0.25%, the oxygen required is approximately 0.12m^3/0.01%C/tonne. The theoretical amount to convert C to CO is 0.093m^3/0.01%C/t. Further oxygen oxidises iron, manganese and to a lesser extent silicon, chromium and phosphorus into the slag. Injection at too high a rate reduces the efficiency of oxidation and increases the oxygen requirement.

For a finishing carbon in the range 0.1-0.2% having removed 0.25% C, the typical quantity of oxygen used would be 5.5-7.0 m³/tonne. It would be usual to remove this amount of carbon in 10-15 min depending to some extent on the furnace size. For a given furnace, the oxygen needed to remove the carbon would be known more precisely than the range quoted above.

Similarly, the probable temperature rise would be known. For furnaces above about 50 tonne capacity, 0.05 m³/t/min will increase bath temperature by about 1°C.

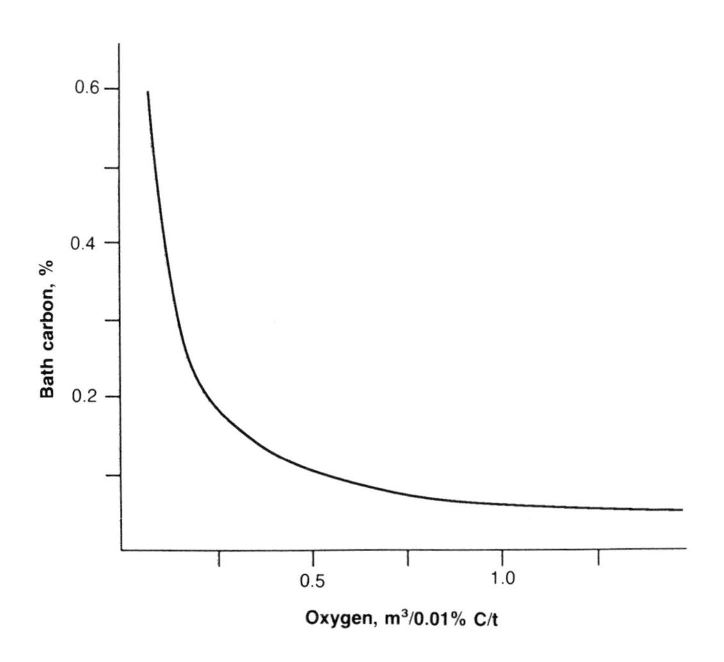

5.2 Practical oxygen requirements for decarburisation as a function of bath carbon

For a 100 tonne furnace where 0.25% has to be removed in 15 min, the oxygen required, its blowing rate and probable temperature rise are given below in Table 5.3.

Table 5.3

Criteria for 100 t arc furnace 0.25%C removal in 15 min

Criteria	Magnitude
Furnace	100 tonnes
Carbon removal	0.25%
Specific oxygen	0.25 m^3/0.1%C/t
Total oxygen	625 m^3
Time	15 min
Oxygen flow rate	41 m^3/min
Specific temp rise	1°C/0.05 m^3/min
Temperature rise	82°C

Desulphurisation

Sulphur can enter the arc furnace bath only from the charge; hence by good scrap management, the sulphur burden can be limited. Sulphur is, however, always present, to some extent, in scrap steel and this is best removed after the oxygen blow which removes carbon and other oxidisable elements. The highly oxidised CaO-FeO "first slag" is removed and a new slag is made by adding calcined lime, fluorspar and carbon to the bath. This is the reducing slag which promotes desulphurisation of the bath.

The sulphur, present in the steel as iron sulphide, reaches equilibrium with iron sulphide in the slag. The iron sulphide in the slag reacts with the lime to form calcium sulphide thus removing iron sulphide from the equilibrium with sulphide in the metal. These reactions are shown below:

$$\underset{\text{(in metal)}}{FeS} = \underset{\text{(in slag)}}{FeS}$$

$$\underset{\text{(in slag)}}{FeS} + \underset{\text{(in slag)}}{CaO} \quad \underset{\text{(in slag)}}{FeO} + \underset{\text{(in slag)}}{CaS}$$

With excess lime (CaO) in the slag, the free iron oxide FeO, which is the oxidising unit in the slag, can be reduced to less than 2%. Such a slag would be very reducing and promote rapid sulphur removal. The small amount of carbon added will reduce the dissolved oxygen content in the metal to the equilibrium level required by bath carbon. This reduction in the metal will influence the oxygen (ie Fe) content of the slag and promote the reducing conditions conducive to sulphur removal.

These can be expressed as:

$$\underset{\text{(in slag)}}{FeO} \rightleftharpoons \underset{\text{(in iron)}}{FeO}$$

$$\underset{\text{(in iron)}}{FeO} + C \rightarrow Fe + CO$$

The small evolution of carbon monoxide prompts gentle stirring which brings unreacted metal into contact with the slag. The fluorspar may contain appreciable quantities of sulphur itself and may affect the slag basicity but physically fluxes and increases the fluidity of the slag which allows the slag reactions to proceed more rapidly and uniformly than would be possible in a viscous slag. Some sulphur is oxidised from the slag into the furnace atmosphere.

It is estimated that reducing slags can remove about 40% of the sulphur in the bath. (Some sulphur is removed by partition into the first oxidising slag) at rates of about 0.006/7%/hour. Such rates are low and 30 to 60 minutes would be needed for desulphurisation under one reducing slag. If sufficient sulphur is removed a second reducing slag would be made after removal of the first sulphur saturated slag. Since such a procedure would add a further 50-60 min to cycle time and hence significantly decrease productivity, it is important to avoid high sulphur charges.

The gentle stirring caused by the carbon monoxide boil can be enhanced to increase the desulphurisation rate. This was originally done .by mechanical rabbling with greenwood poles which also burned to increase gas evolution in the bath still further. Induction stirring and argon gas bubbling also increase sitrring and slag/metal contact and desulphurisation rates of up to 0.009/0.010%/h have been achieved to reach final sulphurs of 0.008/.010%.

More recently, techniques have been developed to inject desulphurising powders with argon or nitrogen into the ladle. Using these methods, which are discussed in Chapter 6, desulphurisation can be completed in about 15 min.

Hydrogen

Hydrogen is introduced into the steel bath by rusted and damp charge materials and additions of lime and fluorspar. Drying of fluxes can be particularly beneficial in reducing the hydrogen load. Some hydrogen is, however, absorbed from the water vapour in the atmosphere. From charging to commencement of the carbon boil at about 1530°C, the hydrogen content of the steel could be in the range 6-8 ml/100g. Oxidising conditions favour reduction of hydrogen and the carbon boil with the evolution of carbon monoxide rapidly reduces the hydrogen content. The violence of the boil directly affects hydrogen removal and this is illustrated in figure 5.3. After the oxygen blow outlined in Table 5.3, ie 1.0%C/h, the hydrogen content could be in the range 1-2 ml/100g.

The danger period for hydrogen pickup comes at the end of the carbon boil when the reducing slag is being employed for sulphur removal. Although the partial pressure of hydrogen is low in the arc furnace atmosphere the arc itself readily dissociates water vapour to molecular and atomic hydrogen.

The latter form of hydrogen diffuses readily into liquid steel where it tends to remain because of the absence of a purging evolution of carbon monoxide during the reducing cycle.

High quality low alloy steels are frequently made in the arc furnace and these steels are particularly susceptible to "hair-line" crack formation caused by hydrogen. For this reason it is usually necessary to vacuum degas these steels to avoid prolonged and expensive hydrogen removal by heat treatment.

Nitrogen

Nitrogen is added to the arc furnace bath in the charge but most is absorbed directly from the atmosphere. The higher the temperature, the greater is the rate of absorption.

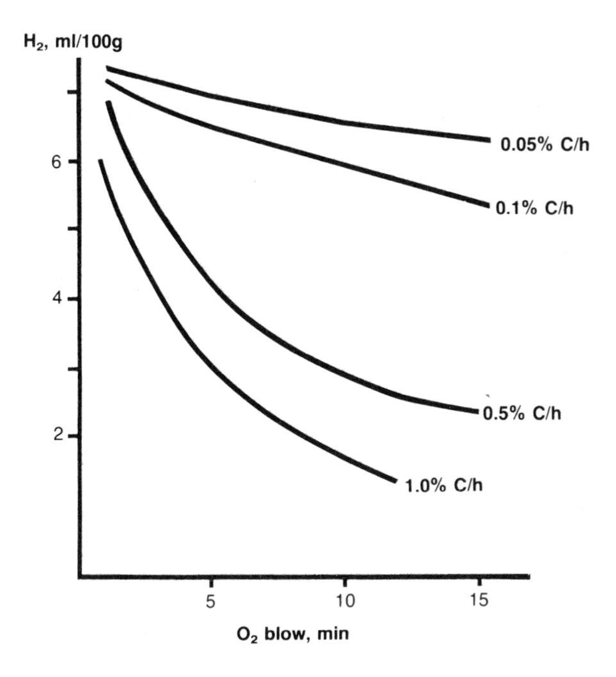

5.3. The effect of carbon removal rate on H_2 removal

Nitrogen is reduced by gas evolution during the carbon boil and if desulphurisation were not necessary (ie single slag practice) then the steel would finish at about 70 ppm nitrogen. Double slag practice (ie one oxidising slag and one reducing slag) produces steels with nitrogen in the range 80-110 ppm.

Deoxidation and final additions

This step would be brought about as for the open hearth by additions of mixed deoxidisers such as ferromanganese, ferro silicon and aluminium. Other ferroalloys to bring the steel into specification would be added and the bath temperature increased to the desired tapping temperature. Trim additions would be made in the ladle.

Residual element control by charge selection

Residual elements cause the greatest problems during reheating and subsequent rolling operations, particularly tin and copper. These elements are enriched at the surface during reheating for rolling by the preferential oxidation of iron, and form low melting point phases with iron. (Nickel, chromium and molybdenum are essential components in alloy steel production, but in many plain carbon steels, these elements are deleterious to properties, for example weldability.) The important feature about residuals such as copper, tin, nickel, molybdenum (and to some extent chromium) is that they are not removed during steelmaking.

The majority of the charge to the arc furnace is merchant scrap bought from outside the steel works. Such scrap has, even within grades, unpredictable variations in physical size, shape and composition. Crushing and baling to increase bulk density to assist charging will retain any non-ferrous scrap and increase copper, tin and aluminium in the steel. Size reduction and magnetic separation will reduce such physical scrap contamination. However, scrap processing is only now becoming accepted and increasing contamination from copper, tin and aluminium significantly affects steel quality.

Such processing cannot segregate different steel qualities. Hence alloy scrap for plain carbon steel charges increases nickel, chromium and molybdenum which will affect end use properties. Free cutting steel (with up to 0.3% sulphur) and cast iron (with up to 2.0% phosphorus) significantly increases the sulphur and phosphorus burden. While these two elements can be reduced during steelmaking, such operations need time and power to bring the steel within specification and the costs of production are thus increased.

The upward trend in metallic residual elements is shown in figure 5.4. At some stage, a high grade of furnace charge, ie one containing less residuals has to be introduced into the charge mix so that the final steel remains within specification. Such iron sources are of higher cost than scrap and are hence kept to a minimum.

One source of high quality scrap of known and consistent composition is virgin scrap directly from a blast furnace - LD steelmaking operation. Such scrap is not usually available on the open market and therefore only available in large steel

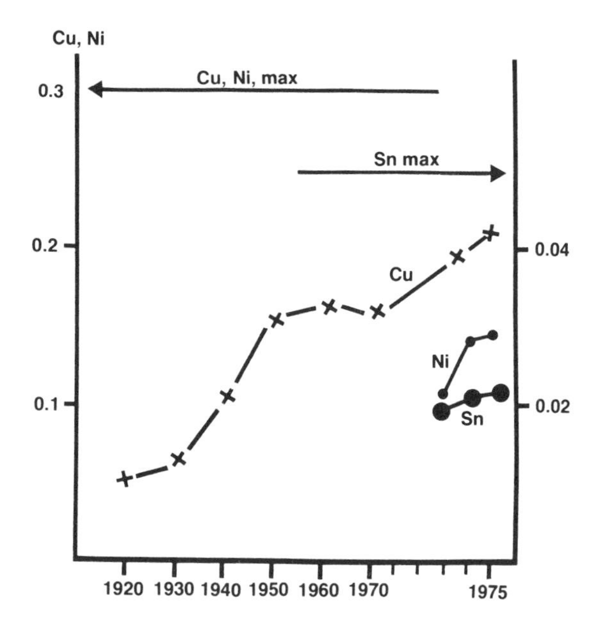

5.4 Upward trend in Cu, Ni and Sn content of steels

conglomerates who operate both arc and BOF shops. Where the residual problem has become acute at the arc shop then BOS scrap would be available.

Another source of predictable consistent quality iron units is blast furnace iron. It is available in broken plate or granulated. Such iron contains at least 4% C whereas most steels from arc furnaces generally contain less than 0.6% C. Clearly there is a limit to the amount of such iron units that can be charged. (The carbon could, of course, be removed by oxygen blowing but this would add time and cost to the operation.)

The growth of the arc furnace as a steelmaking unit using all scrap began to affect the availability of scrap in the mid 1960s. The scrap quality was deteriorating, and its cost to the steelmaker was increasing.

There was, therefore, a need for another source of iron units. Direct reduction of iron ores to metallic iron (not involving the liquid iron stage as in blast furnace operation) had been used on a very limited scale for many years. During the late 1960s and early 1970s several processes using static reactors, vertical shafter kilns or horizontal rotating kilns were developed using gas, oil or coal as the reductant. The product, usually pelletted for ease of handling has predictable and low residual elements and controllable carbon making an ideal charge iron source for arc furnaces. The residual contents of the iron units discussed is given below in Table 5.4.

Table 5.4
Residual contents of various iron units

Source	Cu %	Ni %	Sn %	S %	P %
Pressed & sheared scrap	0.42	0.24	0.02	0.08	0.04
Old baled scrap	0.45	0.10	0.10	0.11	0.04
Mech fragmented scrap	0.13	0.08	0.01	0.03	0.01
Cryogenically fragmented scrap	0.16	0.10	0.01	0.03	0.02
New detinned scrap	0.05	0.01	0.15	0.02	0.01
Turnings (loose)	0.29	0.34	0.02	0.08	0.03
Turnings (briquettes)	0.26	0.29	0.02	0.15	0.03
BOF scrap	0.03/.06	0.02/0.06	0.01/.02	0.01/0.02	0.02/0.04
Blast furnace iron	0.01/.02	0.01/0.03	0.01	0.03/0.05	0.01/0.03
Directly reduced iron	0.005	0.01	0.005	0.01	0.01
Specification	0.3 max	0.3 max	0.05 max	0.06 0.03	0.06[1] 0.03[2]

1. bulk steel 2. alloy steel

It must be stressed that the data given in Table 5.4 for scrap gives only a general indication of residual elements for the

CHAPTER 9
The Production of Oxygen, Nitrogen and Argon

INTRODUCTION

Throughout this book a common theme has been the use of oxygen, nitrogen and argon, singly or in combination in the production of steel. Use varies from 55 m^3 of oxygen/tonne of steel in the BOF process, through the $30m^3$ of oxygen, $20m^3$ of argon and $5m^3$ of nitrogen in the AOD process to $0.03m^3$ of argon per tonne in ladle stirring. The source of all these gases is the atmosphere whose composition is given in Table 9.1.

Table 9.1

Composition of the air by volume

Constituent		Quantity %	Remarks
Nitrogen		78.08	Constant
Oxygen		20.95	Constant
Argon		0.93	Constant
Rare Gases		Balance	Constant
Carbon Dioxide		0.03	Variable-pollution, climate
Water	min	0.02	Variable-climate
	max	5.00	

The three gases range from the very reactive - oxygen, through almost unreactive - nitrogen, to the chemically inert

- argon. Oxygen was the first gas to be identified and separated by Lavoisier and Priestley in the late 1700s by heating oxides of mercury and collecting the product - oxygen. While still available only to the scientific community Bessemer in 1856 in his British Patent 356 was writing "if atmosphere air or oxygen is thus introduced into the metal in sufficient quantities it will . . . . increase its temperature during its transition from the state of crude iron to that of cast steel or malleable iron without application of fuel", and again in British Patent 1292 "it will be understood that pure oxygen gas or a mixture thereof with air or steam may be used". It was to be almost 100 years before basic oxygen steelmaking in a Bessemer like vessel became a reality with the BOF processes.

BRIN'S OXYGEN PROCESS

By 1980, oxygen had become commercially available from the Brin Process which relied on the same principle as in the original isolation but on barium oxide rather than mercury oxide. The process is shown diagrammatically in figure 9.1.

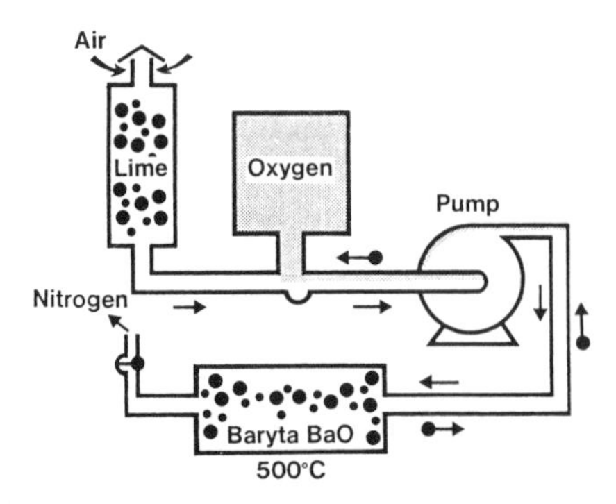

9.1 Brin's oxygen process

The basis for the process is the reversible barium oxide-oxygen reaction:

$$2BaO_2 = 2BaO + O_2$$

Dried and carbon dioxide free air was pumped over barium oxide (BaO) at 10 psi and 600°C when the oxygen in the air forms BaO_2 and the nitrogen enriched air was vented off. When no more oxygen was adsorbed the pumps were stopped and the temperature raised to 875°C when oxygen was liberated. The pumps were reversed and the oxygen drawn off into a storage vessel. 96% O_2 was produced on a 15 min cycle. The main use for the product was for "limelight" - oxygen-hydrogen flame on a block of limestone to produce light for the Victorian lantern slide shows. By 1890, about 25,000 m^3/year of oxygen were used from low pressure bags or heavy cast iron cylinders. The maximum output of the plant was about 10 m^3 per hour.

MODERN AIR SEPARATION PLANTS

Brin plants could not sustain the developing, mainly medical, market but the air distillation system was already being studied by Hampson (Britain), Linde (Germany) and Claude (France). From the early 1900s the air distillation system provided the oxygen and later nitrogen and argon for the world's industry. With considerable improvements in design, operating cycles and manufacture, the system is essentially the same today. A modern ASU (air separation unit) is shown diagramatically in figure 9.2.

Air is drawn into the system through a filter and compressed to about 6 bar by a turbocompressor (C). The compressed air is cooled and washed in the direct cooler (D) and delivered to the reversing heat exchanger (E).

The air is cooled in (E) almost to the point of liquefaction (approximately -170°C). Carbon dioxide and water vapour in the air freeze out and are deposited on the walls of the exchanger. The dry, carbon dioxide free air is then fed to the base of the lower distillation column.

In the lower column, the air is liquefied and separated by distillation into liquid nitrogen and "rich liquid" containing about 40% oxygen (the bottom product). These two liquid

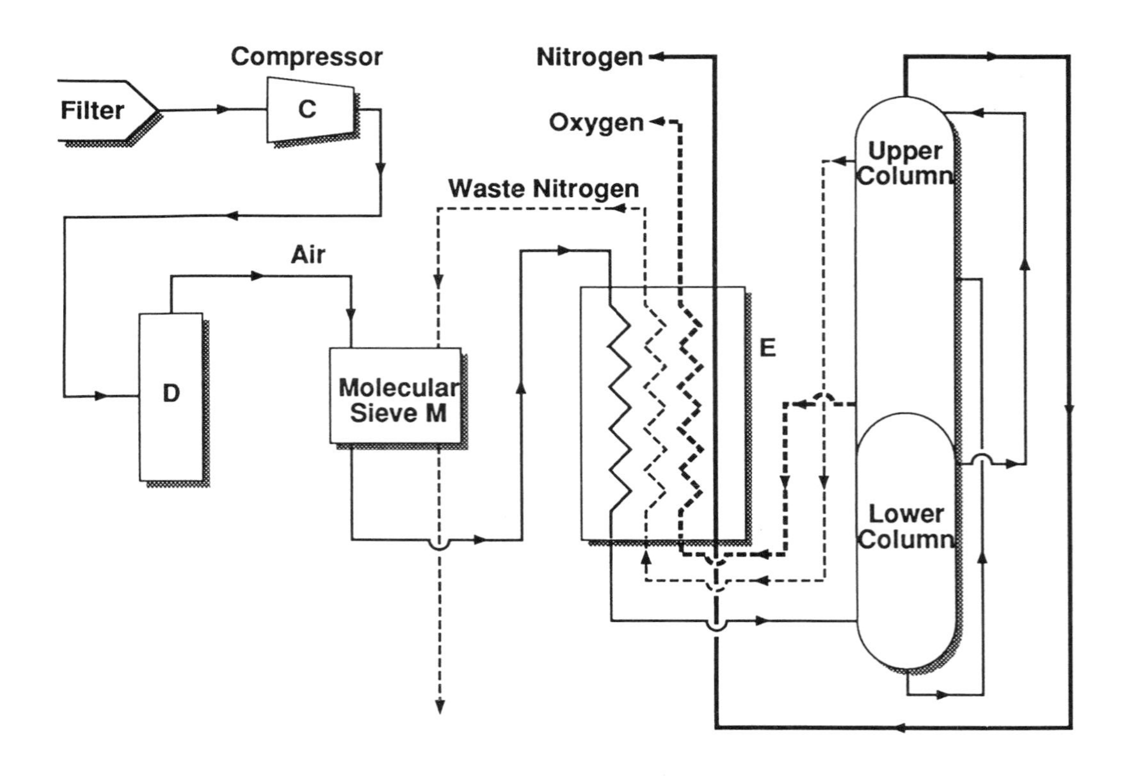

9.2 A modern air separation unit (ASU)

streams are then fed to the upper distillation column where the final distillation of the air into oxygen and nitrogen takes place.

Pure gaseous nitrogen is obtained from the top of the upper column and pure gaseous oxygen is obtained from the bottom.

Waste nitrogen, containing about 2% oxygen, is vented from the upper column at a point several trays below the top. All the gaseous products from the column are returned through separate passages in the reversing heat exchanger (E), cooling the ingoing air, and themselves being warmed to ambient temperature in the process.

The waste nitrogen passage and air passage in (E) are identical, and a special valve arrangement allows the flows to be changed over periodically. The waste nitrogen is thus able to purge the water and carbon dioxide previously deposited from the air, and so prevent the exchanger from becoming blocked.

After leaving (E), the waste nitrogen is vented to the atmosphere, and the product oxygen and nitrogen, which are produced at a little over atmospheric pressure, are compressed to a suitable pressure, or passed direct to the user, as required.

The cold required by the process is produced by isentropic expansion of a portion of the high pressure air in an expansion turbine (not shown in figure 9.2).

The basic low pressure, air separation cycle described above is capable of several modifications. For example, by increasing the cold production capability of the expansion turbine, up to 7% of the oxygen and nitrogen products can be withdrawn from the distillation column as liquids. Larger quantities of liquids can be produced by the addition of a separate liquefier unit. Again by including a side distillation column fed from the main column, crude argon (98% pure) can be produced.

By further liquefaction and distillation very high purity liquid argon can be manufactured from the impure product from the primary plant.

Having made the products, they have to be distributed to the user industry. For steelworks, which are very large users, the producing plant is located adjacent to the works and the

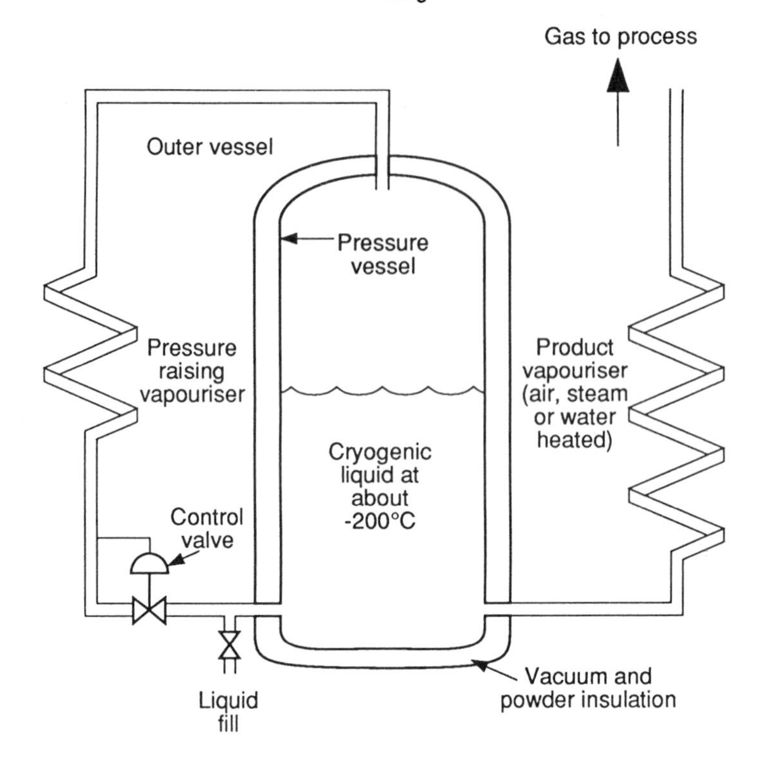

9.3 Liquid gas storage unit (diagramatic)

Table 9.2

Volume equivalent of cryogenic liquids

Element	Volume of liquid (l)	Weight of liquid (kg)	Volume of gas at 15°C and one atmosphere pressure (m^3)
Oxygen	1	1.14	0.840
Nitrogen	1	0.81	0.678
Argon	1	1.39	0.823

oxygen piped at the required pressure and flow to the steelmaking furnaces. In the UK, there are pipelines connecting several major users to oxygen producing plants in Sheffield, the Midlands and the North East. The longest UK pipeline system for oxygen is 27 miles and longer systems exist, mainly in the USA.

SITE STORAGE OF GASES

For smaller demands, up to about 2,000,000 m^3/year, the gases are frequently stored on the user's site as cryogenic liquids in vacuum insulated tanks. The liquids are moved from major plant to distribution centres in 100 tonne vacuum insulated rail cars by British Rail, normally in trains of 10 cars. From distribution centre to users, road tankers, again vacuum insulated, move the cryogenic liquids into site storage. The site storage tanks can hold up to 100,000 litres of cryogenic liquid, and are available in a range of standard sizes.

A typical storage system is shown diagramatically in figure 9.3. The inner vessel contains the cryogenic liquid and the vacuum and powder insulant between the inner and outer vessel reduces heat transfer to the liquid to a very low level. When gas is not being withdrawn, the vessel can stand for about a week before the pressure built up in the tank causes the safety valve to lift and gas to escape. Under normal use, therefore, there is no loss of gas.

When gas is being withdrawn, the pressure is kept constant by the pressure control valve. If the pressure in the system falls below the required setting, the control valve opens and admits liquid to the pressure raising vaporiser. Both this vaporiser, and the product superheater, are of the air heated finned tube type, taking heat direct from the atmosphere.

The capacity of a given storage tank depends on the liquid stored. Table 9.2 gives the gas volume produced from the liquid for oxygen, nitrogen and argon.

CHAPTER 10
The Effects of Technology and Market Changes on Steel Output

The metallurgy of steelmaking processes could be described as an inexact science. It is possible to write down precise chemical equations which explain only in broad terms the complex interaction of elements within the bath. Even with the inexact science, years of experience mean that the product can be specified beforehand and precisely met.

When considering outputs of a plant, country or the world, examination of past production is easy while predicting even the short term future is an even more inexact science. Some common patterns emerge, particularly for technological change, but the effect on the total market varies in both magnitude and time. The factors affecting steel production will therefore be considered first and then applied to predicting production and method of production to the year 2,000.

FACTORS AFFECTING STEEL PRODUCTION

Steelmaking is a very basic industry and after leaving the steelworks the product will pass through several more processes before the true consumer buys the product containing steel. As such consumers buy, orders pass back down the supply line to meet the demand. Eventually there is an order available for raw steel which could be met by several suppliers from several countries. The factors affecting the

choice of supplier are as follows:

	Factor	Contribution to Purchase Decision %
A.	Competitive price (and hence production cost)	25
B.	Suitable for use	25
C.	Dependable delivery	25
D.	Technical service	10
E.	Other	15

The most important are A, B and C, and only when these three factors are about equal is the choice of supplier influenced by second order variations such as technical service and advertising. Secondary influences are more important in steady market conditions.

Various market and technical changes will be considered. It must be emphasised that the comments made on the changes are only a guide as to what usually happens.

STEADY DEMAND

This now rarely encountered condition in the steel industry means that total output from the steelworks is more or less in balance with demand from the direct customers. Individual works would probably have a planned production schedule of four to six weeks and a predictable specification mix. Raw materials needed and semi-finished and finished steel stocks held would be about 10% of annual production. For ease of discussion, a steel industry of 20Mt/year will be considered.

The consuming industries, collectively, would probably carry another 10 to 20% of stocks to ensure smooth production in their own factories.

INCREASED DEMAND

From the steady state more orders become available and a steady increase in demand is identified. Initially, this can be

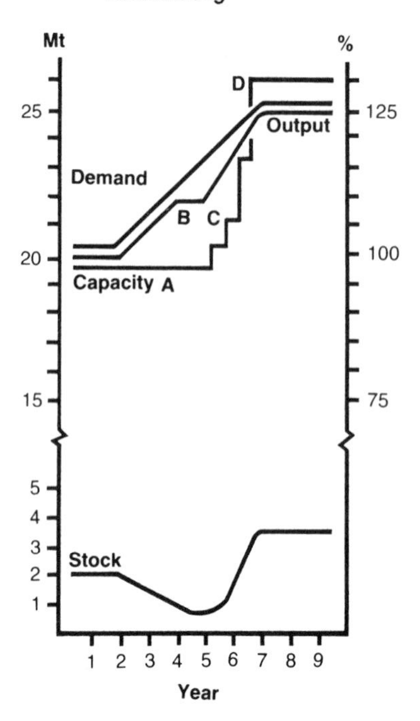

10.1 Increase in demand level

met by reduction of stocks throughout all industry and then by increased output from the existing units in the steel industry. About a 10% increase in demand could be sustained for a reasonable period of time. These changes in levels of operation are illustrated in figure 10.1. The 5% pa growth would be accepted as real within a year (point A) and a need for increased capacity recognised. Steel plants can be from 0.5-5 Mt/year tpy cpacity and take from two to five years to install. At point B imports would begin to rise (as home deliveries become extended) to meet total demand needs. At point C new plant begins to be commissioned, stocks begin to rise and imports reduced. Since most producers wish to supply more steel, excess of capacity is installed in anticipation of the demand continuing to rise. At point D, demand and supply are about in balance, with more capacity available and

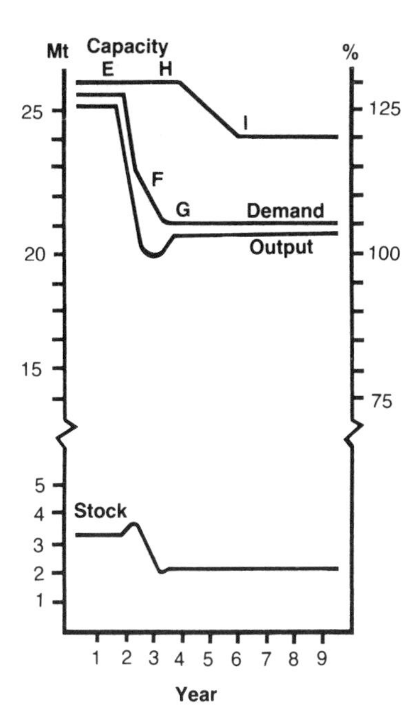

10.2 Decrease in demand level

adequate stocks to deal with the next rise in demand. Generally, the higher the capacity of the industry the more able it is to deal with rising demand.

FALLING DEMAND

Demand can drop significantly and suddenly (point E, figure 10.2) and in such a market condition, users tend to reduce their own stocks and orders at the same time which has a compounding effect on the steelworks. Steelwork stocks rise and output falls. Stocks are then run down and, of course, output has to fall further. At point F, demand stabilises but a further period of time to G is required before demand and output are again in harmony. If demand continued at the low level, some capacity would begin to be shed (H). This would be the older, less efficient, plant generally needing large capital investment for modernising but with little prospect of a return on that investment. Capacity would still be significantly higher than demand because of the expectation (and hope) that demand would again rise.

TECHNOLOGICAL CHANGE

These changes will occur at any time relative to market changes. Generally, during rising demand, emphasis will be placed on a search for methods of increased output and during falling demand on methods of reducing production cost. Enhanced properties and quality can create an entirely new market to add to the existing market or takeover some of a market of other products. There are two types of change - evolutionary and revolutionary.

Evolutionary Change

Improvements to existing processes are evolutionary since the basic method of production has not changed. Once proven, they can be easily incorporated at modest cost into other systems and will improve output or reduce costs by a factor almost always less than two and more usually in the 1:1 to 1:1.5 range. Taken over a whole industry, the market share for a given process is not significantly altered by a single evolutionary change. In times of rising demand, such changes are first to be incorporated and usually retained during falling demand because they have, by that time, become standard techniques. Such changes are illustrated in figure 10.3. Examples of evolutionary changes would be larger vessels; increase in oxygen blowing rate on BOF; higher electrical capacity arc furnaces.

Revolutionary Change

New techniques or processes not currently used by the steel industry cause revolutionary changes which have more than a factor of 2 effect on output. It requires considerable capital investment and risk for the first taker of the new processes who can derive a large advantage over competitors on cost, suitability and or delivery. Having seen the advantages, more units are installed at the expense of the established competing

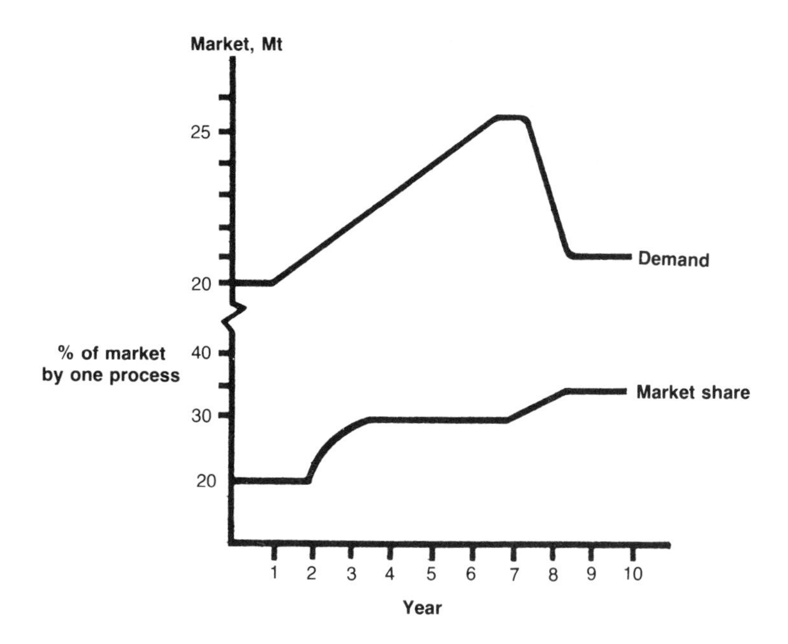

10.3 Evolutionary technological change

process until a natural market share is established. At this peak, evolutionary changes in it and other processes will compete to hold this share for a time. It is inevitable that the share will decrease as other processes are adopted. The growth from idea to established process is shown in figure 10.4. The main points to note are that it usually takes several years for an idea to become an accepted process and by the time it has been adopted, other processes soon begin to reduce its market share.

Examples of revolutionary processes are:

1. BOF oxygen steelmaking

2. AOD stainless steelmaking

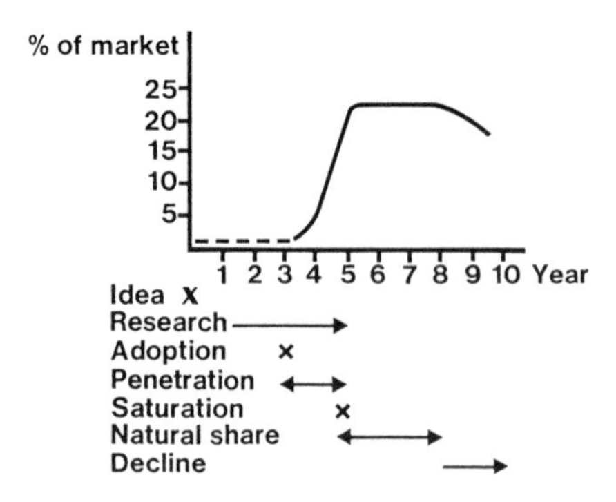

10.4 Revolutionary technological change

but it should be remembered that most or all processes were at one stage revolutionary - even the open hearth.

THE FORECASTING OF STEEL PRODUCTION AND MANUFACTURING PROCESS TO THE YEAR 2000

The real world of steel production is affected by all the mentioned changes simultaneously and the trend is determined by the predominant cause which will vary with plant, company and country. From the complexity of influences and statistics, demand, irrespective of process, will be considered and then the possible share by process.

World Demand

A forecast can be made by extrapolating past trends from available data into the future on the assumption that the trend continues for the period of the forecast. It is worthwhile to consider the trends in production, particularly since 1950, and these are shown in figure 10.5.

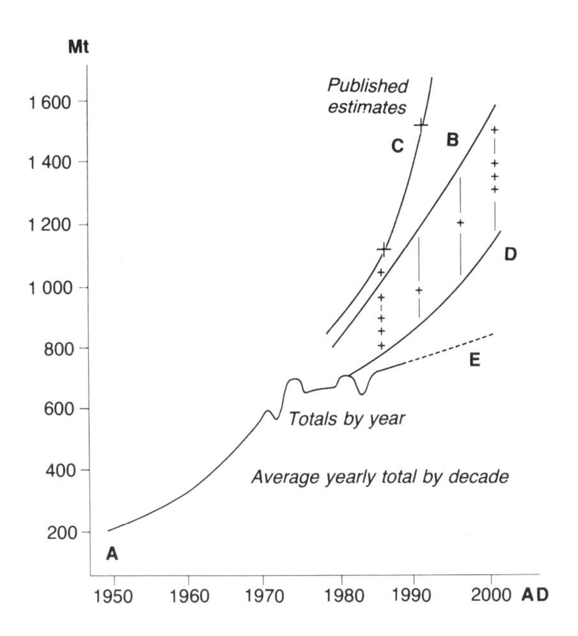

10.5 World steel production 1950-2000.

The history of growth in demand is shown by yearly output at five yearly intervals (A) and for the last 20 years by year. The five yearly interval gives a good trend for extrapolation which would be curve B. While the yearly figures from 1965 to 1973 give trend C.

The data from 1973 to 1982 give trend E, ie, 800 Mt by 2000. The overall trend from 1973 to 1988 is D, ie 1100 Mt by 2000. The + symbols in this diagram are published forecasts made in the late 1970s. It can be seen that all except one are optimistic. The author's forecast is that world steel production will increase to 850 Mt in 1990; 900 Mt in 1995; and reach 1000 Mt in 2000. Growth will not be steady and the change from year to year will be in the range 20-80 Mt - up or down. To put this range in context, the UK total production in 1986 was 14.7 Mt and for the USA 73.7 Mt. With random changes of this order, movement from surplus to shortage on a world or country basis is a fact of life for the steel industry.

Investment in plant, which takes about 5 years to come "on stream" needs high utilisation to pay for that investment. "On stream" may coincide with surplus and cause serious financial problems.

It is such effects that have caused the world industry to change from plant sizes up to 5 Mt considered in the 1960 and 1970 to the 0.5-1 Mt units for installation in the next decade.

The common factor in most of the forecasts is that steel demand will grow and this is based on the assumption that as any nation moves into an industrial economy, its steel consumption grows. Generally, population also grows and with increased use per head, steel demand grows even further.

The 1988 position of examples of developed and developing economics is given in Table 10.1.

Table 10.1

Steel consumption per capita 1988*

Area	Consumption kg/capita
Japan	700
USSR	580
USA	450
Av Western Industrialised	410
EC	380
Average World	155
China	60
Developing Countries	40

*Source: International Iron and Steel Institute

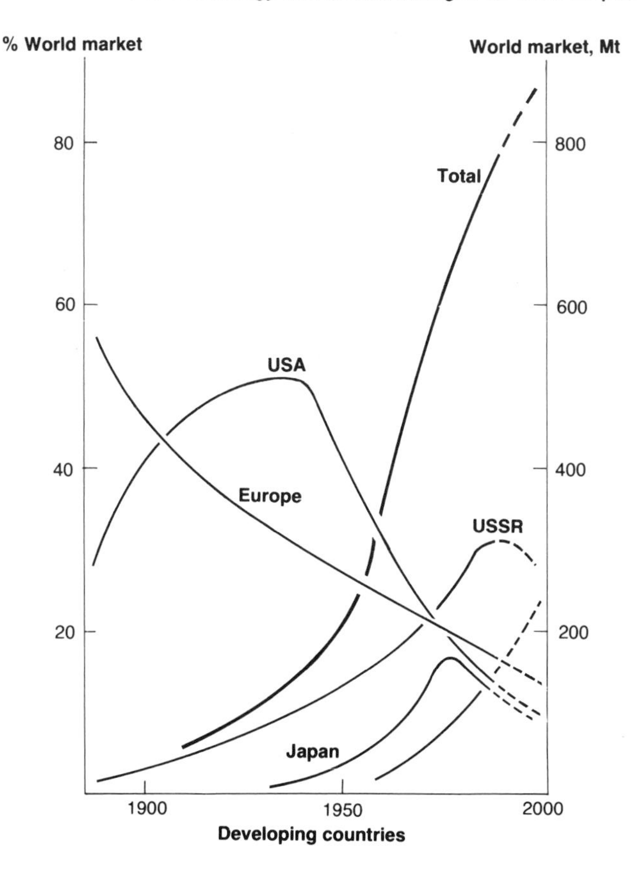

10.6 Relative change in output by developed and developing countries

Generally, a developing country will install more steel capacity pro rata than a developed country to meet its own needs rather than import. The consequence of this is a reduction in the proportion of steel made by developed countries and an increase from developing countries. This trend is shown in fig 10.6. In the late 1800s, Europe was the developed area and USA a developing nation. While Europe grew, USA grew faster until the 1940s. At that time, Japan and USSR grew while Europe and USA grew at a slower rate.

In the 1980s Japan and the USSR have a declining share of world market. The new group of developing nations includes South Korea and China. Capacity reduction in the developed nation has been very significant in the 1980s. The UK has, for example, approximately halved steelmaking capacity since 1979.

Share by Process - Basic Steelmaking

During the 1950s, world primary steelmaking was dominated by the open hearth process which accounted for more than 70% of output. In the early 1950s, increase in output was being met by installation of open hearth shops and arc furnace plants with the Bessemer process declining. This is shown in figure 10.7 where year to year data has been "smoothed". The year data for the 1980s is given to show the year on year variations that do occur.

At the end of the 1950s, the BOF system had been accepted and was being installed to meet further increases in demand. During the 1960s, the rapid decline of the open hearth being replaced by new BOF shops was soon established with a steady but unspectacular increase in arc furnace capacity.

These trends continued into the 1970s and 1980s with the growth rate of arc furnace plants increasing and of BOF plants decreasing.

These changes can probably both be related to a decline in the availability of metallurgical coke for the blast furnace which provides the hot metal for the BOF. With a limit on coking coals, there has to be a limit on BOF capacity.

This has led to a search for alternative commercial processes for reducing iron ore to iron. The direct reduction processes, using fuels other than coking coals have been developed. There are the static reactor type, the vertical kiln type and the rotating kiln system.The two former types of process use gaseous reductants which can be natural gas or synthetic gases made from low grade coal. The latter process uses low grade coals directly. The product is a cold solid product which is an ideal charge for arc furnaces.

A second factor for the growth of arc furnaces is probably the future availability of electric power as opposed to suitable

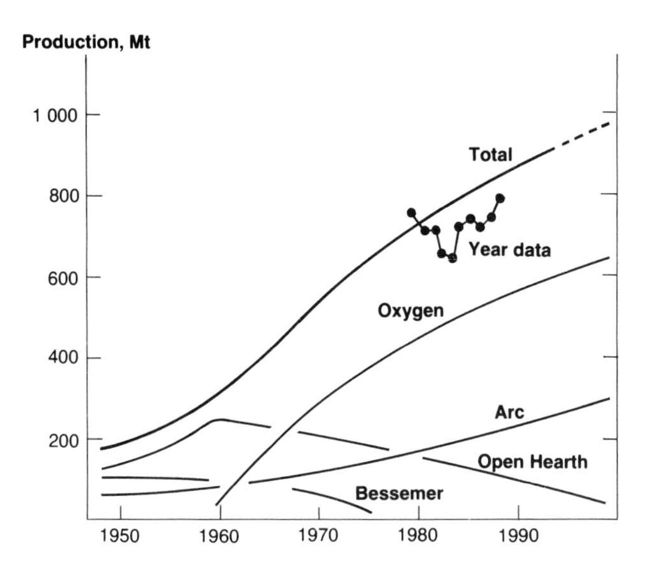

10.7 World steelmaking process changes and growth 1950-2000.

high grade hydrocarbon fuels for metallurgical uses. The low grade fuel can always be converted to electric power.

From these observations, the rate of growth of BOF steelmaking has declined in the 1980s with arc furnace capacity growth holding steady. By the next century it is possible that the arc furnace capacity may approach that of BOF. It is also probable by 2000 that some new process (possibly the plasma process) may be beginning to establish itself in steelmaking.

As can be seen from figure 10.7, changes up to about 1950 were relatively slow. The Bessemer process had been used for almost a century and the open hearth for about 60 years. Changes since that date have been much more rapid. Even so, the BOF will have taken about 40 years from the plant trials in Austria to natural saturation of market share. The arc furnace may take even more time to reach its peak of natural market share. The main reason for these long time cycles is the life expectancy of plant required by the large capital investment involved and because of the variation from the general trend. Longer than forecast "pay back" periods are caused by low demand at the time the plant comes on stream.

Share by Process - AOD

Probably the most spectacular change since the 1950s has been the adoption of the AOD process of secondary refining for the manufacture of stainless steel. This is shown in Figure 10.8. World stainless steel use is generally about 1% of bulk steel and the total market curve has been extrapolated on this basis. In one decade, from 1968 to 1978, the AOD capacity

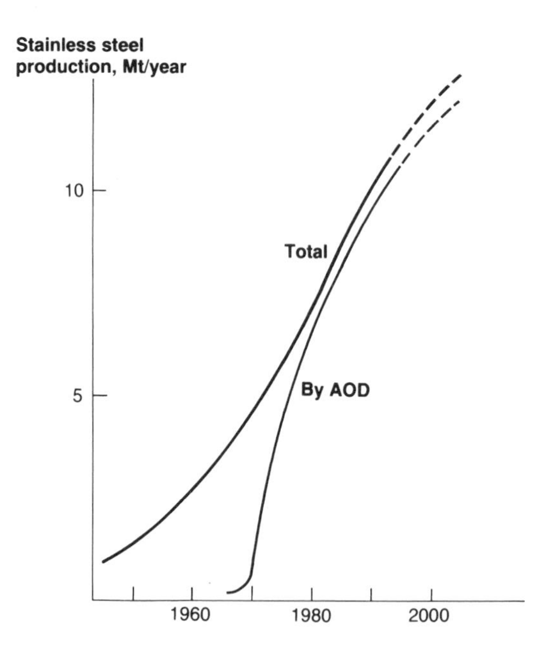

10.8 Growth of world AOD capacity 1950-2000

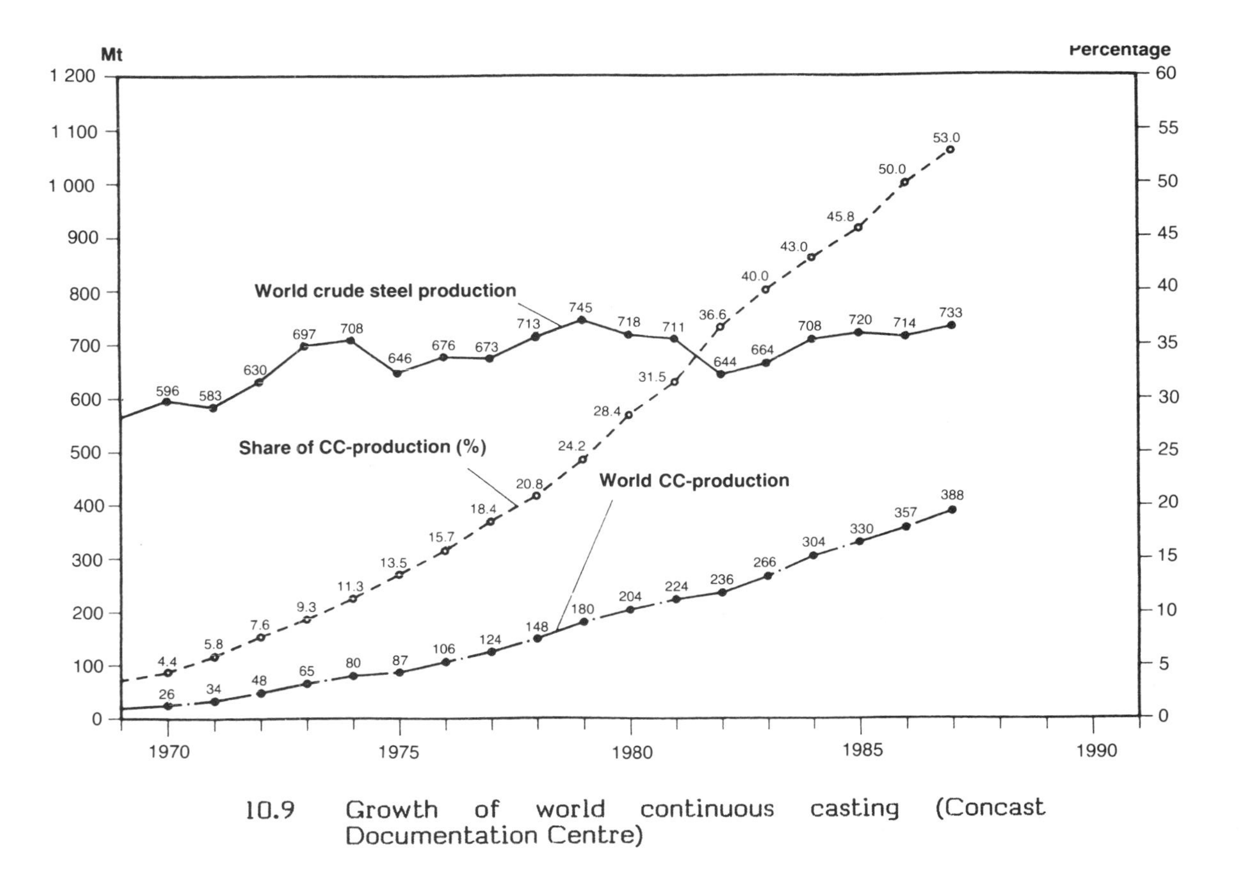

10.9 Growth of world continuous casting (Concast Documentation Centre)

reached about 80% of total world demand for stainless steels. This percentage has risen to over 90% and may now decline a little as other methods (such as vacuum systems) develop and improve. It is unlikely that AOD technology will stand still and the "final" share of the market will be decided by the effect on production cost and quality of any changes made.

Share by Process - Continuous Casting

Continuous casting can be applied to almost all qualities, sections and melting units. It was developed in the late 1940s with the first vertical production plant being installed at Barrow Steel Works in England 1952. The first curved machine for production was installed in 1964 at Von Moos, Lucerne in Switzerland. By 1987, there were 2294 billet machines, 1291 bloom and 543 slab giving a total of 1396 machines worldwide. The growth of steel processed by continuous casting is shown in figure 10.9, the data for which was provided by Concast Documentation Centre, Zurich.

It is significant to note that even though steel production has been stagnant for more than a decade, the amount of steel processed by continuous casting has increased from 124 Mt or 18.4% of production in 1977 to 388 Mt or 53.0% of production in 1987.

Forecasting

In this chapter, prediction as single lines or bands can give an impression of accuracy and certainty as to capacity and market shares. These diagrams do not have that authority. They are an estimate of the future based on information available at the time of the prediction. In general, the predictions for short times are more accurate than for long periods so that the estimate of world steel production in 2000 of 1,400 Mt is probably accurate to ± 20%.

Market share predictions by country or process are less reliable. For instance, predictions of stainless steel production in 1965 to 1980 would show manufacture by arc alone and some vacuum processes. The introduction of AOD in 1968 made that prediction totally wrong. There are probably similar revolutions in process technology being conceived or developed now to make the reality in the future significantly different from what it is thought it should be.

class of scrap to which the analysis is ascribed. Scrap sampling is notoriously difficult as is its "quality control". The analyses do highlight the problem of the arc furnace steelmaker who has to charge large tonnages for different grades of scrap with large variations of analyses within grades. The scrap grades are diluted with known composition iron sources to ensure an adequate safety margin below specification for the steel being made. This is usually 0.1% for copper and nickel, 0.02% for tin and phosphorus. A greater latitude exists for sulphur (probably 0.03%) since this can be removed by furnace or ladle treatment.

Pressed and sheared (not a UK classification) or baled scrap is widely available. Further upgraded scrap is not. Some mechanical fragmentisers are in operation but there is only one cryofragmentiser in Europe and no hot briquetting plant.

THE ENGINEERING OF THE ARC FURNACE SYSTEM

The engineering system will be examined in relation to the metallurgical stages of operation which are given below in Table 5.5. This table has been constructed for the previously used example of a 100 tonne furnace tapping in 4 hours.

Table 5.5 *
Stages in arc furnace melting

Stage	Operation	Duration (min)
1	Charging	20
2	Melt down	120
3	Refining	15
4	Desulphurisation	45
5	Alloying	10
6	Heating to tapping temperature	10
7	Tapping	5
8	Fettling	15
Total	Tap to Tap	240

* This table identifies the stages where engineering changes will have most effect on the overall cycle time. These are

melt down and desulphurisation but small reductions in other stages, cumulatively, can also have an effect.

Initial Capacity

For a new arc shop, the installation of the largest capacity furnace, suitable for production plans, is the first step to take. Since the 1950s, the maximum arc furnace size has increased from about 100 tonnes to about 400 tonnes. For existing furnaces, productivity can be increased by several methods: charge preparation, increase in power input, assisted melting and reduction in metallurgical load.

Charge preparation and handling

Scrap A significant portion of scrap for arc furnaces comes from the "consumer durable" sector, ie scrap cars,

Table 5.6

Bulk density and metallic yields of scrap grades

	Scrap type	Bulk Density kg/m³	Melting Yield %
	Pressed and sheared	600	65-70
	Turnings	600-1200	70-75
1.	N° 5 bales	1300-2000	75
2. 4.	Mech fragmented N° 5 bales*	1000	90-92
	Bale fragments*	2000-2500	90-92
3. 4.	Cryogenically fragmented N° 5*	2000-2,500	90-95
	Cold pressed turnings*	4000	90-95
	Hot pressed turnings*	6000	95-98

* Trial quantities only.

refrigerators etc, which has a low bulk density. The furnace has a finite volume and increasing bulk scrap density could decrease the number of baskets typically used from 3 to 2. Generally, yield of iron from the scrap also increases with increase in density. This is due to the removal of non metallic material and decrease of oxidation during melting. Table 5.6 gives typical bulk densities and yields on melting the range of scrap available.

Directly reduced iron Because the reduction process is a gas-solid reaction the degree of reduction of the iron ore and the carbon content can be controlled. The iron ore, although pretreated, still contains other oxides, eg, alumina, silica and lime which persist to a limited extent in the pellet. The

---

1. Black bales 0.6 x 0.4 x 0.4m of old steel.
2. Mechanical fragmentation usually of crushed car body scrap is produced by mechanically shredding the car body at ambient temperature. Non metallics and non ferrous contaminants are reduced and crumpled pieces about the size of a man's hand result.
3. Cryogenic fragmentation can be carried out on already baled scrap (say N° 5). The bales are reduced to below the ductilebrittle transition temperature by liquid nitrogen and then disintegrated in a hammer mill. Flat scrap about 25mm long is produced which enables a better physical separation of contaminants to be achieved compared with mechanical shredding. Processing costs are higher.
4. The products from both fragmentising processes can be added to the baskets or charged continuously but wire can jam feeding mechanisms. Fragmented scrap is relatively expensive and has higher than normal residuals..

acceptable range of parameters for DRI for arc furnaces is given below in Table 5.7.

Table 5.7

Direct reduced iron for arc furnaces

Parameter	Range %*	Aim
Metallisation	90-96	as high as possible
Total iron	88-94	not more than 2% below metallisation
Metallic iron	83-91	not more than 5% below total iron
Gangue	4-10	as low as possible
Sulphur and phosphorus	0.03-0.05	as low as possible
Carbon	0.8-1.5	dependent on arc furnace practice
Size	> 5 mm	reduce dust burden on fume cleaner

*The ranges given are to cover most arc furnaces, operation. For a given furnace, the actual specification required from the DRI plant would depend on furnace size, steels to be made, quality of available scrap and method of operation, the overall aim being to reduce total production costs. The effect of the variations of the parameters on other aspects of furnace operation (and hence costs) are discussed below.

Degree of metallisation All reduced materials contain oxides of iron which, on melting, will enter the slag phase. Yield losses may be higher due to increased slag bulk but since some pellets have lower gangue content than some scrap, this may not occur. There is no evidence to suggest any difference in slag chemistry between pellet and scrap charge, but any higher oxide contents can be reduced.

$$FeO, Fe_3O_4, Fe_2O_3 + carbon + heat \rightarrow Fe + CO$$

In the first case for wüstite (FeO) it has been shown for an 85 tonne furnace that the energy requirements increase by 10 kWh/tonne per 1% decrease in the degree of metallisation. For the magnetite (Fe_3O_4) and hematite (Fe_2O_3) reactions, the energy increase is likely to be 20 kWh/tonne/1% decrease in metallisation. It is obvious that the metallic content of the pellet should be as high as possible.

Gangue content The gangue, usually silica (SiO_2), can alter steelmaking by changing both slag weight and chemistry. For an increase of 100 kg of slag weight per tonne of steel, there is an increase in energy needs of between 85 and 150 kWh/tonne. It follows that gangue content should be kept to a minimum.

Composition is also important since, if all the gangue content is acidic, ie, silica, then lime will have to be added separately to the furnace to give the desired lime/silica balance. The gangue composition can be modified in the reduction unit to produce a self fluxing pellet. The gangue of such a pellet would have a lime/silica ratio of between 2:1 and 3:1. No additions other than the pellets would be needed.

Generally an all scrap melt will produce between 50 and 75 kg slag/tonne of steel. The gangue content of pellets would have to be as low as 2% to give a similar slag bulk for 100% pellet charge. The desired range of 4-6% for 100% charge would give slag bulks of about 100-150kg. The more usual practice is to mix charge thus reducing the effect of the gangue content on slag bulk.

Charging Practice - Baskets

Most arc furnaces have swing-aside roofs so that the full cross section is available for charging. Scrap from the arc furnace stockyard is packed in baskets or buckets with a clam shell base closure. For the first basket into an empty furnace, light scrap is placed in the base (to act as a cushion) with heavier scrap above and light scrap above again to allow the arcs to penetrate quickly and save roof refractories. The basket is lowered into the furnace and the rope, which holds the base closed, burns through and discharges the scrap into the furnace. The roof is swung back and power switched on to melt the charge.

For a 100 tonne furnace, it would be usual to use no more than three baskets, probably 40, 40 and 20 tonne to complete the furnace charge. Time would be saved and production increased by two basket charging with higher bulk density scrap.

Charging Practice - Continuous

Processed scrap, ie mechanically or cryogenically fragmentised and pellets, all have a consistency of size and shape which make these forms of iron units suitable for bunkering and conveying and hence continuous charging into an arc furnace. Continuous charging can be used to supplement batch charging or used as the only method of charging. In the developing countries continuous charging of pellets is a major production route.

Where continuous charging is required, the iron is fed into a molten heel of metal. This practice has the advantage of decreasing power fluctuation associated with all scrap melting during the melt-down period. The stable melt conditions can reduce power consumption by up to 10% and decrease charging time.

ENERGY INPUT

The prime source of energy is electrical which is transmitted to the steel via the three carbon electrodes. Power is needed not only to melt down the steel scrap and superheat the molten metal, but also to compensate for heat losses to the furnace structure and waste gas system. Melt down time can be reduced by increasing the power available and by supplementary firing with gas or oil burners with or without oxygen. Oxygen alone - burning both the carbon monoxide generated from the charge and electrodes and the light scrap - will also reduce melt down time. These latter techniques are generally known as assisted melting.

Electric Power

The power from the electricity supply company is routed to the furnace through step down, regulating and furnace transformers. The furnace power is available on modern units

at 150V to 600V in 25V steps. In the 1950s it was usual to have installed power of 150-250 kVa/tonne and this has risen to 400-600 kVA/tonne in the 1970s. Such furnaces are known as ultra high power furnaces or UHP. The effect of such increases in power on melt-down time is shown in Table 5.8.

Table 5.8

Effect of power increase on melt down time for a 100 t nominal furnace

Class	Power Transformer MVA	kVA/t	Melt down time h
RP	20	200	3.0
HP	45	450	1.5
UHP	60	600	1.0

RP - regular power
HP - high power

With the higher power, electrode consumption increases and for a UHP would be in the 5.0-5.5 kg/tonne range while in an RP it would be 4.0-4.5 kg/tonne. Because of the significant decrease in melting time, even at a high power input, it is possible to reduce specific power consumption from 580-620 kWh/tonne to 500-520 kWh/tonne.

Assisted Melting

Air-fuel burners have been used to preheat scrap in the scrap baskets. This technique will remove water from scrap and prevent explosions in the furnace from trapped water and reduce hydrogen pick-up. It is usual to heat the scrap to between 200 and 500°C. At the higher temperatures, the efficiency of heat transfer falls and heat losses increase so the process becomes less economical. At high preheat temperatures, welding and bridging of scrap near the burner can lead to problems in discharging the basket into the furnace.

The most advanced use of air fuel burner techniques is in the SKF MR process which uses a double shell furnace. Scrap is charged to one shell and preheated by an oil burner. The burner and roof are exchanged for an electrode roof so that melting can be completed. The use of two shells and interchangeable roof also permits two melting cycles - suitably out of phase - to be carried out at the same time. Residual heat in the furnace after tapping helps to preheat the scrap and since the scrap is already in the melting unit, high preheat temperatures are possible.

There are three basic techniques used in assisted melting which are in the order of supplementary energy added:

1. Oxygen infiltration without a hydrocarbon fuel.

2. Oxy-fuel burners with burner fired substoichiometrically.

3. Oxy-fuel burners with burners fired super stoichiometrically.

Oxygen Infiltration

The waste gases from the arc furnace contain between 15 and 30% of carbon monoxide and hydrogen. These gases are not normally burned in the furnace but by gentle lancing of oxygen into the scrap above the liquid bath, these gases can be burned and the heat of the reaction released in the furnace. Carbon can be added to the charge and light scrap will also be oxidised to release heat. The reactions involved are:

$$2CO + O_2 = 2CO_2 + \text{heat}$$

$$2H_2 + O_2 = 2H_2O + \text{heat}$$

$$C + O_2 = CO_2 + \text{heat}$$

$$2Fe + O_2 = 2FeO + \text{heat}$$

Typically 10 m^3 of O_2/t would be used with 5-10 kg carbon/t saving 10-15% of tap to tap time and 50 kWh/t of electric power.

Oxy-fuel burners (substoichiometric firing)

Using burners, the rate and amount of supplementary energy can be controlled provided that the waste gas system can deal

with the increase in volume and temperature of the extra gases produced.

To distribute the energy release, up to 3 burners would be used especially on furnaces exceeding about 60 tonnes capacity. The burners would be individually mounted to fire through special ports in the side wall so that direct impingement of flame on the electrodes and side walls is avoided.

The burners used are controllable both in total heat release and the ratio of oxygen to fuel so that desired stoichiometric requirements are used. The stoichiometric ratio controls both flame temperature and free oxygen in the waste gases. The relationship is shown in figure 5.5. For practical purposes, flames with only 1% free oxygen are neutral and at this level oxidation of the charge due to the burner would be negligible. Maximum flame temperature occurs at about 85% oxygen where there would be about 7% free oxygen in the flame. This level would be the maximum with some oxidation of the charge. The actual operation condition would normally be set for a particular furnace operation and for plain carbon steels this would be 75%.

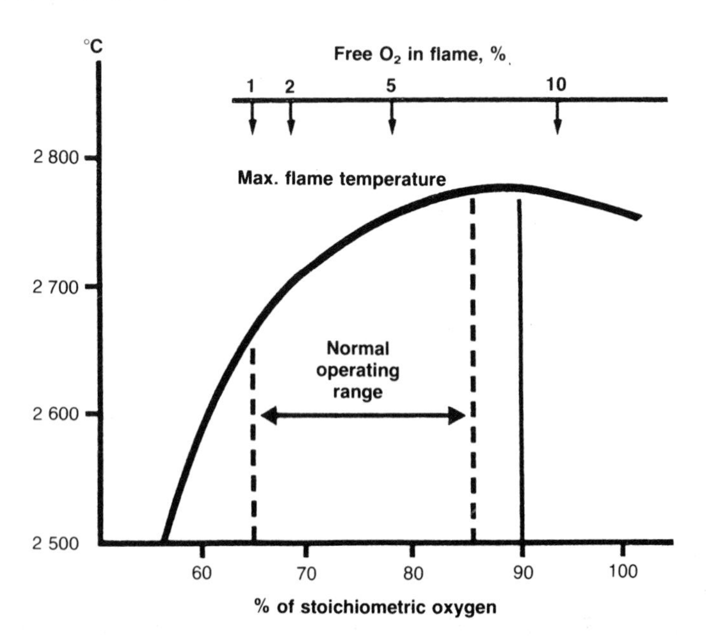

5.5 Relationship between stoichiometric ratio, flame temperature and free O_2 for O_2 and natural gas

It is usual to fire the burners only while solid scrap remains so that there is an adequate heat sink for the thermal energy released. The burners are therefore turned off when 50% - 60% of scrap has been melted. This mode of practice would use no more than 28 m^3 of oxygen/tonne on a small furnace and 15 m^3 of oxygen/tonne on a large furnace. The corresponding fuel input rates would be about 220 kWh/t (20 l oil/t) and 110 kWh/t (10 l oil/t).

This method of substoichiometric firing can be used for almost any grade of steel including low and medium alloy steels because the level of oxidation of the charge is controlled and minimised. The supplementary thermal energy can be used to reduce electrical power without increasing melting rate by operating at lower transformer settings thereby reducing maximum demand charges. Such operation is aimed at reducing melting costs and depends on the relative costs of electric power, oxygen and fuel.

It is more common to use the technique to reduce melt down time and increase output. The additional cost of fuel and oxygen is then compared with alternative means of increasing output and the extra profit from the higher sales. The time and power savings possible for a range of furnaces is given below in Table 5.9.

Table 5.9

Savings with oxy-fuel assisted melting*

Furnace size t	5-10	20-25	70-80	100-120
MVA rating	2-3.5	8-10	15-20	40-60
Fuel kWh/t	220	150	140	100
Oxygen m^3/t	28	21	17	15
Time saved min	40	30	30	20
Power saved kWh/t	120	120	100	60

*Electrical energy is saved by using heat energy from the oxy-fuel burner and time savings of 25% are typical. With the reduced operation time, total heat losses and hence energy are also decreased.

Oxy-fuel burners (Superstoichiometric Firing)

Aimed at maximising arc furnace output for bulk and low alloy steels, this technique was developed at the Toshin Steel Co in Japan. The technique was developed on a 50 tonne furnace and applied on a 140 tonne furnace and is used by many other operators throughout the world.

Three oxy-oil burners capable of being fired at 200% stoichiometric ratio are used for about 70% of the melt down time with the furnace at full electrical power. The oil contributes about 60 kWh/t but the excess oxygen by burning scrap, carbon and metalloids from ferroalloy additions contributes a further 145 kWh/t. The burners are supplemented by oxygen lances which not only cut the scrap down into the bath but also infiltrate oxygen into the furnace to burn carbon monoxide and hydrogen. With correct metallurgical control, the 50 tonne furnace had a tap to tap time of 1 h 8 min and the 140 tonne, 1 h 27 min giving a production rate of 100 t/h.

The increase in output from one 50 t furnace as the technique was developed is shown in Fig 5.6, showing the combination of

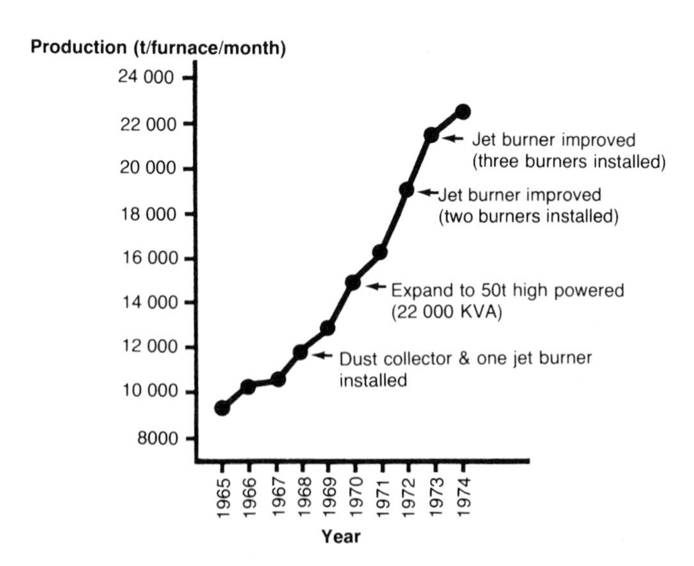

5.6 Increase in output of 50t furnace as Toshin System developed.

electrical and thermal energy. The details of one cycle are given below in Table 5.10 and the use of consumables and production rates achieved are given in Table 5.11.

Table 5.10

Details of furnace operation (Toshin System)

Operation	Actions	Duration (min)
Repair and charge	1st basket in	6
Melt down	2nd and 3rd baskets in Burners on - 35 min Lances on - 44 min Carbon 0.10-0.25% Slag off at 1540°C	48
Refining	Decarburising, Dephosphurising Increase temperature to ~ 1620°C Lances on for 6 min Slag off	11
Finishing	Desulphurising to 0.04% max Killing and final additions	3
Heat time	Tap to tap	68

These very high production rates which the oxy-fuel techniques make possible can only be achieved when the engineering of the system can contain the energy release involved. The two most significant changes are the split shell to enable the side walls to be removed completely from the furnace at sill level to permit relining of the furnace. This means that one shell can be removed and a relined shell repositioned with a minimum of delay.

The second innovation is the introduction of water cooled carbonaceous brick panels in the side walls. With these and mechanised relining equipment, well maintained high speed

ancillary equipment such as cranes, improvement in electrode quality, balancing of gas output and fume extraction to minimise air in leakage and training of skilled operators makes such short tap to tap times possible.

Table 5.11

Operational results from sequence given in Table 5.10

Quantity	Dimensions	Value
Liquid metal yield	%	91.1
Tap to tap time	min	68.0
Productivity	t/h	41.8
Consumables		
Power	kWh/t	333.0
Oxygen	Nm^3/t	42.3
Oil	litres/t	6.1
Electrodes	kg/t	4.4
Wall refractory	kg/t	2.1
Roof refractory	kg/t	3.4
Ferro alloys	kg/t	18.8

Materials handling

The yearly amounts to be handled depend on the arc furnace practice used. The lowest quantities are required by an alloy steelmaking furnace using oxygen for decarburising only, followed by substoichiometric oxy-fuel firing with the highest output from a Toshin type practice. The amounts required for 18 shift per week operation for 45 weeks per year is given in Table 5.12 for a 100 t arc furnace.

FUTURE DEVELOPMENTS

Already being applied on a small number of arc furnaces, is the use of coal as a supplementary fuel. Electric power consumptions for electric melting alone are in the range 450-520 kWh/t. Power use as low as 300 kWh/t when injecting

Table 5.12

Products and Consumables for 100 t arc furnace

Quantity	Oxygen for Decarburisation	Sub-Stoichiometric Assisted Melting	Super-Stoichiometric Assisted Melting
Production rate			
t/h	25	33	66
tpy	162,000	214,000	428,000
Tap to tap			
min	240	180	90
Liquid metal			
yield %	95-96	95-96	91-92
Scrap	170,000	225,000	465,000
Electric power			
kWh/t	600	540	333
Oxygen m^3/t	6	20	42
Oil l/t	-	9	6

25-30 Kg of coal/t and combusting. This with oxygen are probable. Oxygen use increases to about 35 m^3/t for the coal combustion.

Gas injection through annular tuyeres in the base of the unit is also being tried. The tuyeres are similar to those used in AOD vessels, ie there is a methane or an inert gas in the outer annulus with inert gases and or oxygen in the centre pipe. The gases injected provide bath agitation and are intended to promote rapid reactions and reduce melt time. The geometry of the arc furnace, does not, however, help in the uniform promotion of reactions and this development is less likely to succeed than coal injection with oxygen.

CHAPTER 6
Secondary Steelmaking for Stainless Steel (AOD, CLU)

INTRODUCTION

In this section the AOD (argon oxygen decarburisation) and the CLU (Creusot-Loire-Uddeholm) processes are covered. In terms of commercial adoption, there are about 100 AOD vessels ranging from 3 to 150 tonnes and only 2 CLU units of 70 tonnes and 30 tonnes. This chapter therefore concentrates on the AOD process. Processes for the manufacture of stainless steel will be described. In both cases the metal has been melted in an arc furnace, as described in Chapter 5 and then transferred to the secondary refining vessel, ie both vessels are converters. Vacuum manufacture of special and stainless steel is dealt with in Chapter 7.

To explain the AOD and CLU processes it will be necessary to outline stainless steel production from an arc furnace alone so as to emphasize the importance of the vessel and explain the metallurgical reactions.

STAINLESS STEEL PRODUCTION FROM THE ARC FURNACE ALONE

In the arc furnace, decarburisation is brought about by lancing the bath with pure oxygen. The product of decarburisation is CO essentially at 1 atm pressure and the stainless steel is in equilibrium with that gas at the temperature of the bath. These conditions determine the maximum amount of Cr that can remain in the bath at a given carbon content and temperature. The equilibrium curves are shown in figure 6.1.

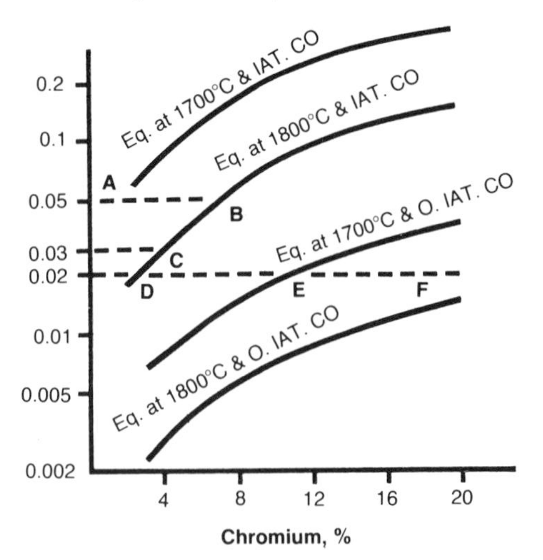

6.1 Carbon-chromium equilibrium curves

The curves clearly show that, as the carbon is removed from the bath, the equilibrium chromium content is also lowered and any chromium in excess of equilibrium is very rapidly oxidised out. If the temperature is raised, then the amount of chromium that can remain in the bath is also raised. For example, at 0.05% C and 1700°C (point A, figure 6.1) the equilibrium chromium content is approximately 2%. At 1800°C (point B) the equilibrium chromium is approximately 7.5%.

Most stainless steel production requires carbon at or below 0.03% at which the equilibrium chromium level, even at 1800°C is only 4% (point/C). Since carbon is contained in small amounts in all ferro alloys, for a specification of 0.03%C it would be necessary to finish at 0.02%C to allow for added carbon even where low carbon ferroalloys were used. At this level of carbon the equilibrium chromium is about 3% (point D).

In conventional arc practice, for the manufacture of an 18% Cr 8% Ni 0.04%C steel, it was only possible to use a charge mix with 4% Cr in it. The molten charge would be blown down

with pure oxygen to about 0.02% C giving a temperature of about 1800°C. During the blow about 2% of the charge chromium would be oxidised into the slag. About 1% would be reduced back into the metal by using ferrosilicon giving a bath analysis of 3% Cr, 0.02%C for a specification of 18% Cr and 0.04% C. (The nickel content is largely unaffected by the decarburisation reaction).

Large amounts of ferrochromium of the very low carbon type were added to bring the bath analysis to 18% Cr and still remain under 0.03% C. These large amounts of alloys also cooled the bath to the required tapping temperature.

The natural penalties of the arc only process are summarised below:

1. Low charge chromium only possible. This meant that only small amounts of stainless steel scrap could be used.

2. Very high temperatures - 1900°C - in the furnace requiring high cost refractories.

3. Large additions of high cost, low carbon ferrochrome.

4. A lower limit on carbon of about 0.025%.

THE AOD PROCESS (Union Carbide Corporation)

The AOD process was developed from a laboratory study of methods of producing the data given in figure 6.1. When blowing small melts with oxygen, it was not possible to maintain isothermal conditions because of the very exothermic reaction of chromium oxidation. It was to control this reaction that argon was used to dilute the oxygen and so reduce the rate of chromium oxidation and the attendant heat release. The technique worked and it was observed that for a given temperature and carbon level, the chromium that could remain in equilibrium was significantly increased. It was concluded that the partial pressure of CO in the decarburisation reaction had been reduced and the data obtained, produced the two lower curves in figure 6.1 for a partial pressure of CO of 0.1 atm.

Using the 0.02% obtained at 1800°C (point D) where only 3% Cr could be retained in the bath, it is now possible to retain

11% Cr at 1700°C (point E) and 18% Cr at about 1750°C (point F). The research work was started in 1954 and progressed through larger laboratory trials to a 15 tonne pilot industrial scale in 1968. In 1978, the installed AOD vessels had a production capacity approaching 5 Mt/year.

The AOD technique is to melt the charge in the arc furnace with all the elements added in the lowest cost form, ie stainless steel scrap and high carbon ferroalloys. At melt out, the chromium is up to 0.5% above the minimum in the specification to be made with carbon in the range 0.25%-2.0% and silicon 0.2% to 1.5%. The particular levels of carbon and silicon depend on the specification, practice and vessel size as does the temperature of transfer which is about 1500°C. A simplified sequence is given in Table 6.1.

Table 6.1

AOD Sequence & composition changes

Composition %

Action	Cr	Ni	C	Si	S	Temp°C	Ar/O_2	Time (min)
1 Aim at	18.0	10.0	0.025	0.5	0.015	1550	-	-
2 Charge AOD	18.5	9.8	1.5	0.3	0.035	1500	-	2
3 Blow 1	17.8	10.0	0.4	0.1	0.03	1700	1:3	25
4 Blow 2	17.0	10.1	0.15	-	0.03	1720	1:2	15
5 Blow 3	16.5	10.1	0.018	-	0.03	1740	2:1	15
6 Reduce	18.2	9.9	0.02	0.5	0.02	1650	Ar*	5
7 Desulphurise	18.2	9.9	0.02	0.5	0.01	1625	Ar*	10†
8 Trim	18.2	10.1	.02	0.5	0.01	1550	Ar*	1

* Ar stirring of bath
† Includes slag off

The main features to note are the progressive dilution of the oxygen with argon to control both temperature rise and partial pressure of carbon monoxide in the bath. The consequence of that control is the retention in the bath of almost all the charge chromium. Return to the melt from the slag of

oxidised chromium is achieved by ferrochromium silicon and desulphurisation by lime.

The Metallurgy of the AOD Process

Carbon and Chromium

The heart of the process is the control of the partial pressure of carbon monoxide to control the rate of the oxidation of carbon and chromium so that chromium is retained in the bath while carbon is removed to low levels. To show the interaction of chromium and carbon during AOD operation, the chemical changes given in the AOD operation of Table 6.1 have been combined with the theoretical equilibrium data from Figure 6.1. The combined data is given in figure 6.2 where both AOD and arc only changes are shown.

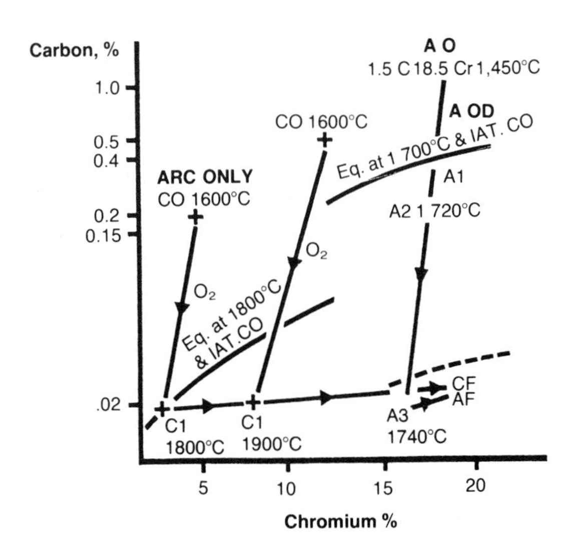

6.2 Composition changes in stainless steel making.

The vessel is charged at 1500°C with 1.5% C and 18.5% Cr (AO). The initial blowing gas is 75% O_2 and 25% Ar. Such a mixture promotes rapid oxidation of carbon but at the low temperature and high C of the bath only a small dilution effect is required. Towards the end of the blow as the carbon content falls below 0.5% then the dilution effect of the argon on the partial pressure of the carbon monoxide is beginning to have an effect. Blow 1 is stopped at 0.4% C, 17.8% Cr and a temperature of 1700°C which is approximately at the equilibrium point for a 1 atm pressure of CO (A1). If the temperature had been higher than 1700°C then scrap coolant would have been added to reduce the temperature to about this level prior to beginning Blow 2. For Blow 2 higher Ar dilution of the oxygen is needed to reduce the partial pressure of CO to promote carbon removal without significant temperature increases or Cr loss. This blow is stopped at 17% Cr 0.15% C at a temperature of 1720°C (A2).

The final blow is 65% Ar 35% O_2 so that the partial pressure of the carbon monoxide is approaching 0.1 atm. Carbon removal continues with a small temperature rise to give bath conditions of 16.5% Cr, 0.018% C at 1740°C (A3). 2% Cr has been oxidised into the slag so ferrosilicon and lime are added and gently stirred with argon alone. About 1.7% Cr is reduced back from the slag to bring the steel into specification with regard to chromium (AF).

In contrast, two arc furnace only practices have been outlined on the same diagram. The starting point of oxygen only blowing is designated C0 and the end of oxygen blowing C1. The right hand C0-C1 curve represents a high charge chromium melt (14% Cr and 0.5% C) which has to be blown to about 1900°C to achieve 0.02% C losing 7% Cr to the slag in the process. This practice represents lower charge costs but high refractory costs in the arc.

The left hand curve shows a low charge chrome practice (4% Cr 0.2% C) which reaches 1800°C to achieve 0.018% C. There is less wear on arc refractories but a high cost penalty for low carbon ferrochrome to bring the steel up to the 18% chromium required (CF).

In both cases, unit costs would be higher than the AOD and the final carbon would be about 0.025%. It is difficult and expensive to achieve lower carbons than this with arc only practice. Steels with carbon of less than 0.01% have been made in AOD units.

Sulphur

In arc furnace practice, desulphuring by rabbling with a lime slag is time consuming. One or more treatments requiring more than one deslagging operation can be needed to get the sulphur down to levels below 0.02% (depending of course on the sulphur load in the charge).

In the converter, even with high charge sulphurs of 0.04%-0.06%, sulphur contents down to 0.005% have been achieved in a short desulphurising step after the chromium reduction (step which also removed some sulphur) by adding desulphurisers and stirring with argon.

Lead

From some grades of scrap the lead content in the bath can reach 0.010% at which level there are serious problems during the rolling of the product leading to low yields of saleable product. The use of the AOD with normal control of leaded scrap probably giving melt out lead contents of about 0.007% can reduce the lead to the 0.001-0.002% range where no problems are experienced in rolling.

Oxygen

When oxygen alone is used to decarburise, oxygen can reach 0.35% requiring significant amounts of deoxidant to reduce the oxygen content to an acceptable level in the range 0.005-0.010% (50-100 ppm). This reduction produces oxides which may not be completely removed from the system.

In the AOD, oxygen contents at the end of the third blow can be in the range 0.15-0.25% reducing to 50-100 ppm during the argon only stirring for reduction. The reduced amount of deoxidant needed to remove the oxygen in the AOD steel produces a lower volume of oxides and hence cleaner steel. Oxygen contents in the semi finished product were reported as being 20% lower than when made by the AOD route compared with arc only production.

Nitrogen

In the simplified sequence given in Table 6.1, the gases used were oxygen and argon. Nitrogen can be used as a

replacement for argon depending on the stainless steel being made. For low nitrogen melts, nitrogen will be used as a process gas up to the end of Blow 2 but thereafter argon must be used. The purging effect of the argon reduces the nitrogen picked up during the earlier blows down to the 100-350 ppm range. Where high nitrogen contents are required, then nitrogen can be used as a process gas in Stage 3. In these cases, nitrogen contents in the range 1,000-2,500 ppm can be achieved. These levels are usually required for high proof stress alloys where the alternative method of increasing the nitrogen content is to add high cost nitrided ferroalloys.

Hydrogen

Hydrogen contents in the range 3-5 ml/100g are typical of AOD practice. Levels down to 1 ml/100g have been observed as have contents up to 7 ml/100g. The high contents (5-7 ml/100g) usually result from partially hydrated lime used in the reduction of desulphurisation steps. These levels can be reduced to the normal range by argon blowing for 1-2 minutes prior to tapping.

THE ENGINEERING OF THE AOD SYSTEM

With every AOD vessel there is the associated arc furnace of comparable capacity to provide the hot metal charge. The arc will not be considered in this section, only the blowing vessel itself. To show the magnitude of the engineering problem, the materials to be handled for a 50 tonne capacity vessel are given in Table 6.2.

The Vessel

A typical vessel is shown in figure 6.3. Capacities range from 5-100 tonnes with 2-6 tuyeres on the back wall so that when the vessel tilts to tap or deslag, the tuyeres are lifted clear of the metal. The tuyeres can have an upward tilt up to 5° and are generally within an angle of 120° symmetrically along the back wall, and about 20 cm from the floor of the vessel.

The tuyere height is approximately constant because metal depths are no greater than 1.5m and usually between 1 and 1.5m. Specific volume, ie vessel volume to steel volume is in

the range 0.7-0.9 m^3 per ton. It is interesting to note this is the same range for the BOF vessel shown in figure 4.6. Both vessels are gas-steel reaction vessels but with different injection techniques which change bath depth needs but do not change the reaction volume.

The refractory lining of the vessel has to contain a very aggressive metallurgical reaction and thicknesses from 0.3-0.4m are needed. Even so the lining life is generally about 40-50 heats with some vessels reaching 80. The severity of the wear can be highlighted by comparing the life of an LD vessel lining which has a typical range of 300-500 heats and over 1,000 heats possible.

Since wear is severe, there are large variations from any average figure which is about 25 kg/tonne of steel. Lining wear adds to the slag volume but has no effect on refining chemistry which is a gas-metal reaction.

Table 6.2

Consumables & products of a 50 t AOD vessel

Hot metal from arc		93,000	tpy
Scrap		5,000	tpy
Ferro Alloys		4,000	tpy
Lime etc		9,000	tpy
Refractories		2,500	tpy
Oxygen	30 m^3/t	3,000,000	m^3/year
Argon	20 m^3/t	2,000,000	m^3/year
Nitrogen	5 m^3/t	500,000	m^3/year
Stainless steel		100,000	tpy
Slag		13,500	tpy
Gas		6,000,000	m^3/year

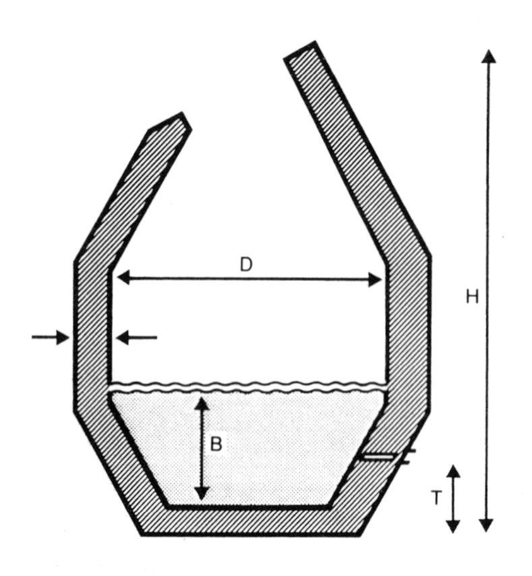

Parameter		Symbol	Ratio	Example
Capacity	t	C		50
Volume	m^3	V	V/C (.7)	28
Height	m	H	H/D	5.1
Diam	m	D	H/D (2)	2.7
Bath	m	B	H/B (3.5)	1.5
Tuyere	m	T	B/T (7)	0.2
Lining	m	t		0.35

6.3 The AOD vessel

Process Gas System

Total gas blowing rates are in the range 0.7-1.0 m^3/t to give adequate stirring of the bath without ejection of metal from the vessel. The gas system has to supply this rate which for a 50 tonne vessel would be 3,000 m^3/h. The mixture changes through the process as detailed in Table 6.3.

Table 6.3

Gas changes in AOD operations

Condition	Composition %			Flow rate	Duration
	O_2	Ar	N_2	m^3/min	min
Blow 1	75	-	25	50	25
Blow 2	34	66	-	50	15
Blow 3	66	34	-	40	15
Reduction	-	100	-	15	5
Tuyere Shroud	-	100	-	4	60

Supply pressure of all gases is up to 16 bar and the control system changes the composition of the mixture as the operator requires. The tuyeres are made from concentric tubes and down the outer annulus flows pure argon to cool and protect the inner tube down which the oxidising mixture passes. Average use of the gases is as follows:

Oxygen	30 m^3/t
Argon	20 m^3/t
Nitrogen	5 m^3/t

The O_2 needs for a given steel can be derived from:

$$O_2 = 2 + 9.3C + 8.0Si + 1.4Mn + \frac{17.2}{1 + 50\,C_1}$$

where O_2 = m^3/t
C = initial carbon in melt %
C_1 = required final carbon %
Si = initial Si + Si for reduction
Mn = initial Mn + Mn from reduction

Nitrogen and argon depend on the nitrogen specification in the steel and the amount of oxygen to be blown. Generally the total inert gas is somewhat less than the total oxygen.

Output and Cost

Since the AOD vessel is a secondary refining vessel linked to an arc furnace of the same capacity, output from the original plant can be doubled. One heat can be processed in the AOD whilst the next charge is being prepared in the arc. With a slightly shorter tap to tap time in the AOD, the process can be easily integrated and synchronised to achieve double output. Standing charges can therefore be carried by the higher output.

However, another vessel requires added capital but the charges for a refining vessel are lower than charges for another arc furnace.

Operationally, the costs in arc furnaces have been reduced since the arc acts as a melting unit at lower temperatures than required for refining. The costs at this stage are reduced by lower consumption of electrodes, power, refractories, labour and oxygen. High carbon ferrochrome is used instead of expensive low carbon qualities.

With the added vessel, costs are incurred for argon, nitrogen, oxygen, refractories and labour. The overall balance is in favour of the arc-AOD system which also produces a higher quality product than the arc alone.

THE CLU PROCESS

In this process, the main diluent of the oxygen is steam to control the carbon removal while maintaining high chromium levels in the stainless steel. As for the AOD, the charge is melted in an arc furnace and transferred to the CLU vessel. Transfer carbon up to 3% and silicon up to 1.2% with all the

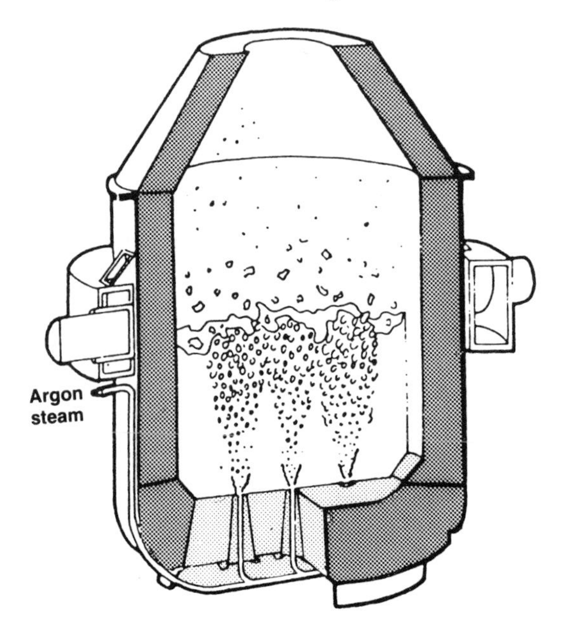

6.4 The CLU vessel

Table 6.4

Process consumables in the CLU process

Consumable	Use per tonne
Steam	10 - 25 m^3
Nitrogen	10 - 25 m^3
Argon	1.5 - 4 m^3
Lime	50 - 80 kg
Silicon	5 - 20 kg
Oxygen[1]	10 - 40 m^3

arc furnace slag has been practised but carbon up to 1.2%, silicon up to 0.5% with some furnace slag is more usual. This practice results in less refractory wear and a lining life of about 85 heats. The vessel is shown in figure 6.4 showing that the process gases are injected from the base.

The process has a 1-1.5 hour blowing time in which control of the steam oxygen and nitrogen in the blowing gas holds the temperature to less than 1680°C, reduces carbon and retains chromium. The use of steam however increases the hydrogen content and from 1.5 to 5 m^3/tonne of argon are needed to purge the hydrogen to less than 1 ppm. The process consumables are given in Table 6.4. The specific use depends on the grade of stainless being produced.

CHAPTER 7
Secondary Steelmaking - Ladle Processes (Stirring, Degassing, Powder Injection, Ladle Steelmaking)

INTRODUCTION

Whilst not, strictly speaking, originally true steelmaking processes, ladle treatments bring about the final adjustments in analysis or temperature which may be small in themselves, but have a profound effect on casting and working within the steelworks and benefits to the end users. Currently, this is the fastest growth area in steelmaking.

The processes which will be dealt with in this section include, atmospheric and vacuum systems for stirring, degassing and analysis adjustment. It is essential in any of these processes for the furnace slag to be removed.

LADLE STIRRING

This class of process is shown in figure 7.1 where a gentle stream of argon or nitrogen bubbles rise through the steel producing an "eye" of hot exposed steel. The steel is gently stirred so reducing any variations in temperature or composition within the ladle. For continuous casting, minimum temperature variations are required for optimum operation so that some form of stirring is essential.

A third effect is that oxides in the steel float upwards and become entrained in the synthetic slag which has been added after the furnace slag has been removed or prevented from entering the ladle. The cleanness of the steel is thus improved for a gas use as low as $0.01 m^3$/tonne to $0.03 m^3$/tonne.

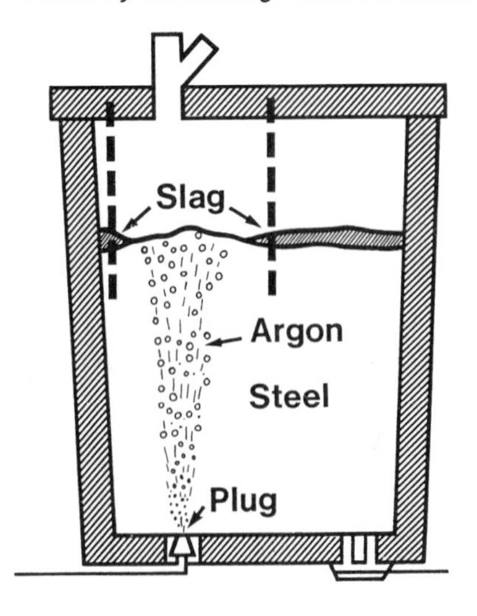

7.1 Ladle stirring

Stirring times rarely exceed 10 min and are more usually 5-8 min since the ladle is losing 2-4°C per minute of treatment time. Some of the heat loss is radiation from the eye where some oxidation also occurs. To avoid this, two methods have been adopted on a limited scale. The SAB (sealed argon bubbling) process uses a refractory box which is immersed over the "eye". Radiation losses are reduced and an argon atmosphere is maintained over the "eye" to prevent oxidation.

A second alternative is CAB (capped argon bubbling) where the same benefits accrue. An alloy chute is usually positioned over the "eye" to facilitate trim additions and give very tight control over composition.

For all these methods, the life of the porous plus is critically important. The conical plug may last the life of the ladle base and, therefore, not require changing during the ladle campaign. It is, however, probable that the plug will wear faster than the ladle base and need replacement. There are several methods of removal from the outside of the ladle base and replacement of the plug would be carried out at the same

time as nozzle block replacements. Lives in the region of 1-15 heats can be obtained depending principally on the grade of plug, steel and casting temperature. Some systems are deliberately replaced after every heat to ensure easy start-up of gas flow.

Stirring may also be done using a hollow stopper rod fitted with a porous refractory cap. This method has the advantage of being independent of the ladle, and gas flow can be checked before immersion in the liquid steel. Should the porous cap fail, then a duplicate rod can be rapidly positioned and used. Changes with full ladles are not possible with base plugs where failure can, in very rare instances, lead to break-out and loss of steel.

In the development stage of this process in the late 1950s and early 1960 s, removal of the dissolved gases, hydrogen, oxygen and nitrogen (when stirring with argon) was investigated. In these cases, inert gas volumes of 0.2, 1.0 and 2-3 m^3/tonne were used. Reduction in hydrogen contents of 3-40% were achieved from initial levels of 2.9-6.2 ml/100 gm. At the lower gas use, O_2, in 80 tonne ladles was reduced 20-30% from initial levels in the 60-100 ppm range and with 3 m^3/tonne reductions of 40 –70% from the 60-80 ppm range were achieved. When argon was used to reduce nitrogen, 0.2-0.3 m^3/argon/tonne reduced nitrogen contents by 2%-15% for the 43-130 ppm level.

Degassing has occurred but the volumes of inert gas used are large. At low flow rates, long process times of up to 30 min and significant temperature drops occur. The temperature fall is approximately doubled by the injection of an inert gas into a ladle and can be as high as 8°C/min. The temperature fall is not principally due to the cooling effect of the gas but due to the radiation losses from the free surface of the ladle. Gentle flow is therefore required and such flow can be defined as that giving a quiet surface with an "eye" of hot metal of about 0.5m diameter. High flows would give turbulence with slag entrained in the flowing steel and hot metal exposed over the whole surface of the ladle.

Because of these problems, bubbling alone is not used to degas but simply to reduce temperature stratification in the ladle and make the metal composition homogeneous. Oxygen reduction does occur by the floating out of solid oxides into the slag. Even when using only 0.1-0.3 m^3 of inert gas

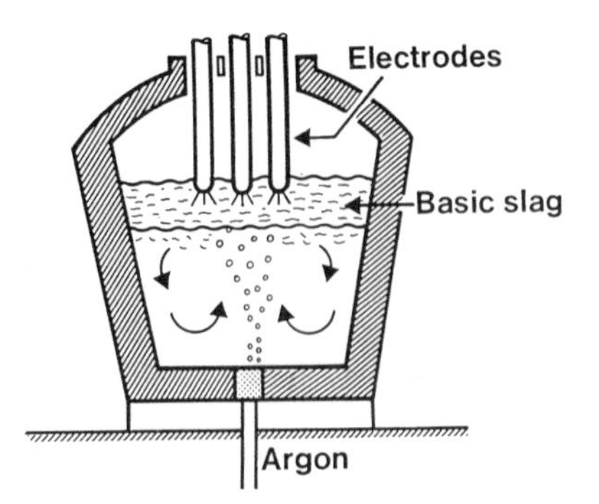

7.2 Ladle furnace

per tonne, it may be necessary to tap the steel some 10°C hotter than normal to compensate for losses during stirring.

Since the restriction on the use of stirring processes is the temperature loss during the treatments it is natural that heating should be added to the system so that not only are higher tapping temperatures not needed but very accurate teeming temperatures can be used. A possible system is shown in figure 7.2.

Argon stirring of the steel is carried out under a basic slag heated by electrodes in a neutral or reducing atmosphere. Oxygen contents of less than 20 ppm and sulphur contents of 0.002% are posssible at an accurate teeming temperature and inclusion size can be reduced to a maximum of about 10μm.

This system coupled to a pneumatic steelmaking unit could produce the special low and medium alloy qualities normally made by an arc furnace at very high output rates.

DEGASSING

Gas stirring alone will not remove hydrogen to an acceptable level in an acceptable time. It had been realised for many years that the application of vacuum to steelmaking could reduce hydrogen - the principal cause of failure in large shaft

and disc castings. Large nitrogen vacuum systems, however, did not become commercially available until the 1950s when both mechanical pumps and steam ejector methods were developed. It was expected that pulling a vacuum above the surface of liquid steel would remove the hydrogen and the ladle vacuum system was probably the first developed. It was found, however, that the steel at the bottom of the ladle remained undegassed because of the ferrostatic head of metal above it. Particular difficulties are encountered in degassing fully killed steels.

At the same time, degassing using argon to remove hydrogen without a vacuum was being investigated.

Such processes depend on bubbles of argon transferring other gases dissolved in the steel to within the argon bubble. The net effect is a reduction of gases in the steel. The theoretical removal of hydrogen and nitrogen by argon bubbling is shown in figure 7.3. The volume of argon required would be such that considerable treatment time would be needed with a

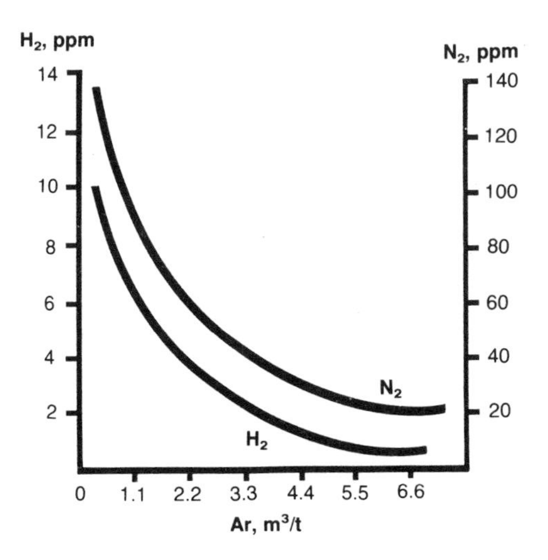

7.3 N_2 and H_2 removal by argon

significantly higher tapping temperature from the furnace to compensate for the cooling during treatment. Clearly, neither this nor simple vacuum systems could be the answer. To effectively and economically remove hydrogen and nitrogen, both vacuum and stirring were necessary. Stirring could be by gas, mechanical or electrical induction.

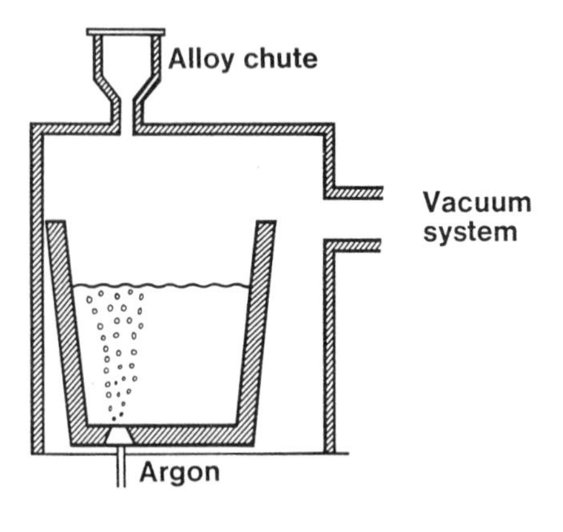

7.4 Ladle inert gas degassing

Ladle degassing with inert gas stirring is shown in figure 7.4. The gas provides metal currents that bring new metal to the surface where the vacuum degases the metal. The Ruhrstahl Heraus (RH) degasser. shown in figure 7.5, uses a twin-legged evacuated vessel to degas steel. The injection of about 0.1 m^3/min of inert gas into one leg reduces the density of the steel which flows upwards into the degassing chamber. When degassed, the steel returns to the ladle via the other leg and circulation of the steel from ladle to vessel ensures complete degassing of the steel.

The Dortmunder Hoerder (DH) process shown in figure 7.6 uses a single legged vacuum chamber which, by vertical up and down movement induces metal flow from ladle to degasser. Modern variations now include inert gas injection to assist in circulation and degassing.

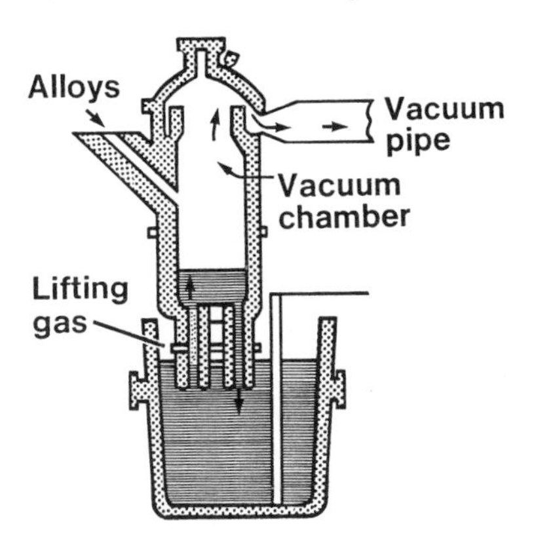

7.5 R H degasser

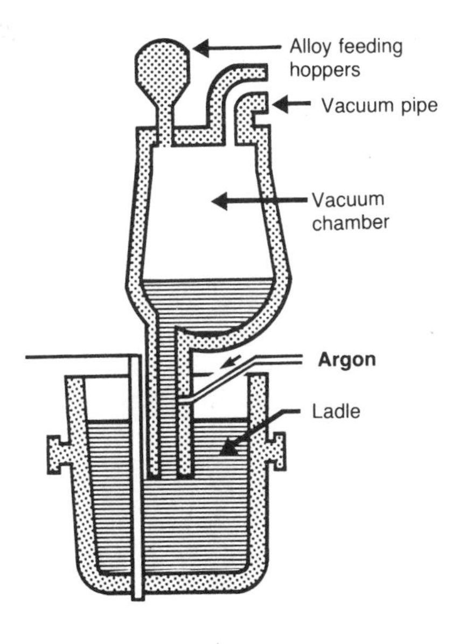

7.6 D H degasser

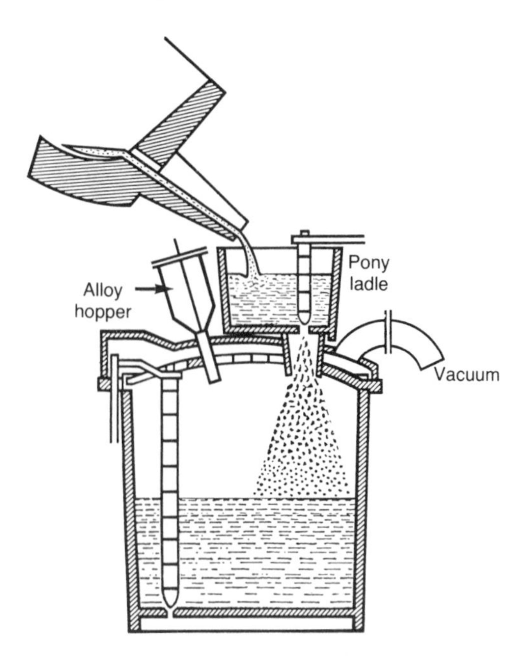

7.7 Tap degassing

Tap or stream degassing, shown in figure 7.7, uses a pony ladle from which the steel is drawn into the tapping ladle by the vacuum. The stream physically breaks up allowing the vacuum access to a large metal surface area so that effective degassing is achieved.

With simple modifications all the above systems can be easily accommodated into existing ladles. When induction stirring is used non magnetic ladles, usually stainless steel, have to be used at much higher cost. It must be emphasised however, that where induction stirring is used, the equipment is capable of more metallurgical control than is possible with ladle, DH, RH or tap degassing. Of the latter degassing processes, DH and RH have been installed by more steelmakers than any other process. Both these processes are capable of reducing hydrogen down to about 1 ml/100g to give high quality steel products.

With all these methods, heat is lost during the degassing process requiring that the steel be tapped hotter from the steelmaking unit thereby incurring higher costs on refractories and energy. Treatment times take 10-20 minutes with up to 3 times the ladle volume passing the DH and RH vessels. Hydrogen removal approached 50% with some decarburisation due to carbon monoxide removal.

Heat losses are lower since the treatment occurs within a smaller sub-vessel while the ladle surface is protected by a synthetic slag. In 300 tonne ladles total temperature losses of only 15°C can be attained. In smaller ladles (say 50 tonne) the loss would approximately double to 30°C. Final hydrogen contents would be in the range 1-3 ppm and carbon, from a starting range of 0.03-0.05% would be about 0.01%-0.015% for a 15 minute cycle. Because of the benefits of degassing on bulk steelmaking, it is estimated that about a quarter of all production is degassed by some method.

For some steels, even more sophisticated processes have been developed where the ladle becomes a furnace and vacuum, heating, oxygen injection or addition lids with gas or induction stirring of the steel can be carried out. Such processes can accurately control composition, (including gas content) and tapping temperatures. Processes which include most of these steps are ASEA-SKF, Finkel-VAD, vacuum oxygen decarburisation (VOD) and others developed by steelworks to suit their special needs.

ADDITION SYSTEMS

Perhaps the oldest of these processes is the Perrin Process developed in 1933 to remove sulphur. Premelted reducing slag was poured into the collecting ladle before the steel was tapped. The violent mixing of steel and reducing slag produced the required desulphurisation and since hot slag was used, temperature losses were acceptable.

About 15kg of reducing slag per tonne of steel was needed to reduce sulphurs to 0.01%. The slag required heating to melt and superheat so the process overall was a very expensive method of desulphurising.

To reduce the costs and time taken during desulphurising, systems capable of injecting powders of high desulphurising

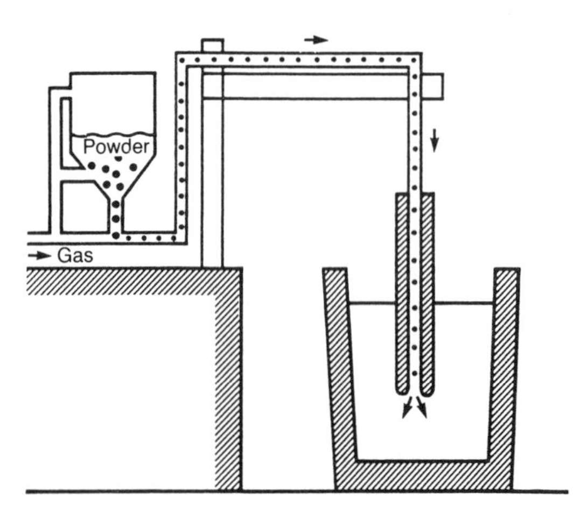

7.8 Ladle injection system

efficiency have been developed. The principles of the system are shown in figure 7.8. The powder to be added is usually raised above the ladle and can be pressurised and agitated by gas to ensure a steady and controllable delivery rate to the lance. The lance is submerged to near the bottom of the ladle to ensure stirring of the whole ladle and a rapid evenly distributed desulphurisation reaction. The gas may be nitrogen or argon depending on the steel being treated and the acceptability of nitrogen being absorbed by the steel.

The volume of carrier gas required is usually in the range 0.1-0.2 m^3/tonne of metal at flow rates of 0.5-1.5 m^3/min for a process time of 8-10 min. The reaction is violent compared with ladle stirring but with the covered ladle reducing heat losses, total temperature drops of about 15°C can be achieved. This can be compensated for by a similar increase in tapping temperature from the steelmaking vessel.

For effective desulphurisation, a low oxygen content in the steel is essential. It is therefore necessary to prevent carrying over the furnace slag and in most cases have a basic lining and slag in the treatment ladle. Suitable compounds would be calcium carbide, calcium silicon or magnesium silicon added at

the rate of 0.75-2.25 kg/tonne depending on the additive and the amount of sulphur to be removed. Sulphurs down to 0.005% are possible using this method.

There are also other significant benefits which occur and are probably attributable to the calcium/magnesium compounds which form. The usual soft manganese sulphides present in steel deform to lenticular shapes on rolling which can initiate or propagate cracks during service use (particularly in tube or plate). With calcium or magnesium treatment the strong affinity of these elements for sulphur and oxygen causes complex small spherical inclusions to be formed. These inclusions are also much harder than manganese sulphide and so do not deform on rolling and resist crack propogation. Very much improved impact properties are therefore obtained in the product.

The difficulty of flowability and nozzle build up and possible blockage with aluminium killed steels is well known. With steels treated with calcium and at high (say 60 ppm) calcium in the steel there is a marked increase in flowability of the steel. This reduces teeming time and hence the temperature fall during ingot or continuous casting. This can affect the need for higher tapping temperatures from the furnace.

The alloy recovery of trim additions of elements with a moderate affinity for oxygen (eg manganese and chromium) can be consistently close to 100% when the alloys are added just before the injection process. This increased yield of high cost ferroalloys may completely offset the costs of the injection process.

Whilst most work has been carried out on the desulphurising process, the powder system could be used:

1. for microalloying with ferroboron, ferrotitanium

2. for recarburisation

3. for adding lead or lead sulphide to free cutting steels

LADLE STEELMAKING

Stirring, degassing and analysis trimming all began as "add-on" processes to the main steel melter and making furnace. Now

Table 7.1

Main operational consumables on

100 t Ladle Furnace

Electricity	40 kWh/t
Electrodes	0.3 kg/t
Argon	100 l/t
Ladle refractories	7.8 kg/t
Treatment time	45-90 min

Table 7.2

Metallurgical control on a 100 tonne ladle furnace

Element	Control Range %	Tapping Value %
C	± 0.015	-
Al	± 0.01	-
Mn and Cr	± 0.05	-
S	< 0.005	0.02-0.04
H_2	1.7 ppm	Degassed
O_2	18 ppm	

the modern ladle process is reducing the function of the arc and BOF furnace to the provider of clean molten steel. This means that the primary making units can be more efficient in that function with a lower metal transfer temperature to complete steelmaking in the ladle. Such units are capable of holding and heating, stirring, degassing and controlling composition for the continuous caster or ingot mould.

A 250 tonne unit with arc heating, argon and electromagnetic stirring, and alloy feed has been installed in a BOF Melting Shop. A 100 tonne unit with a 32 MVA electrode power supply has been installed with an arc furnace. This particular ladle steelmaking, has a vacuum exhaust capacity of 350 kg/h under 1 torr and a stirring capability of up to 15 m^3/h. Typical use of consumables are given in Table 7.1.

The metallurgical control that is possible using the unit is given in Table 7.2.

CHAPTER 8
DC and Plasma Arc Processes (Melting, Ladle and Tundish Heating and Dust Treatment)

DC ARC FURNACES

The engineering experience that has been gained from AC arc furnace operation has been incorporated into the development of DC arc furnaces. This work has been mainly in Russia and France where in about 1980, 12 tonne single electrode units were the largest. Now a 75 tonne, 60 MW unit has been in operation in France since 1984 and units up to 170 tonne, 126 MW are feasible.

Graphite electrodes are used but the number is determined by the furnace capacity and mode of operation rather than the three that are required by AC operation. The arcs are transferred from the graphite electrodes via the scrap charge and molten heel of steel into return anodes in the furnace bottom. There are usually as many anodes in the base as electrodes through the roof. The design, cooling and installation of the base anodes is the critical factor in obtaining good metallurgical control and safe reliable operation of the furnace.

The initial drive to develop these units comes from the lower electrical noise or flicker generated by DC rather than AC furnaces. Because of this phenomenon, high capacity DC furnaces can be connected to grid systems of low short circuit capability at a similar cost to lower powered AC units. The cost of DC rectification equipment, however, is higher than the transformer for AC. It is generally accepted that the DC total installation costs are higher than the comparable AC furnace of the same capacity.

With the limited published operating data from DC units, other improvements over AC are being established. These are:

1. Lower electrode consumption

   In the smaller furnaces with a single electrode, consumptions as low as 1 kg/t have been achieved while 2 kg/t is more typical. With argon atmospheres, this has been further reduced to 0.5 kg/t. These are very significant savings in operational cost compared with the 2.0-3.5 kg/t in AC, 3 electrode furnaces.

2. Reduction in melting power

   30 kWh/tonne can be saved but this is probably, in part, due to the use of a molten heel of about 10% of charge weight being retained in the furnace between heats.

3. Lower refractory wear

   This results from more stable DC arcs and less "flare" and in the case of single electrodes centrally located a greater distance between arc and wall than in 3 electrode AC furnaces.

4. Noise

   Noise reductions of 15dBA down to 92dBA have been recorded on comparable small single DC and AC furnaces. This could avoid the use of "dog houses" for furnaces which have been used on some AC furnaces to meet noise regulations.

The conventional AC and DC arc furnaces use graphite electrodes up to 750 mm in diameter to transfer energy via the arc to the steel charge. Furnace sizes up to 250 tonnes are in operation with most units in the 30-150 capacity. In the larger units, the electrodes are transferring of the order of 50 MVA into the steel change.

PLASMA ARC FURNACES

Introduction

Arc plasmas that are being developed to production status for steel processing today, have evolved from the TIG welding

processes of the 1950s, which used only a few kilowatts of power. In transferred plasma arc torches, the arc is generated between a thoriated tungsten electrode and the charge using a plasma gas which is usually argon.

A plasma is generated when any gas is raised to a temperature at which a large fraction of the gas species present are ionised and will conduct electrical energy. The temperature needed depends on the gas and will be in the range 6,000-20,000°K. Argon is the more usual choice of plasma gas as it is conductive at a low temperature, which helps start up, and it is chemically inert and so protects the electrode from oxidation. Where the plasma is a DC system, the plasma torch is usually the cathode because a lower energy is released at cathodes compared to anodes. The cathode thus requires less cooling and the charge benefits from the higher heat release.

However, AC and anode torches have been developed and with suitable design and water cooling are used in specific conditions.

One form of the DC cathodic plasma torch is shown in figure 8.1.

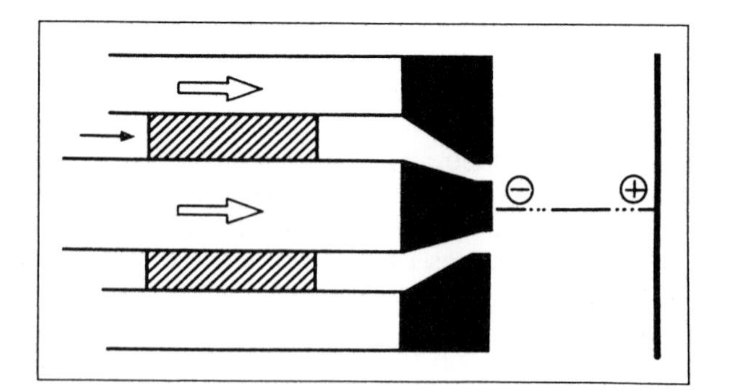

8.1 The cathode plasma torch

The nozzle and cathode, with its thoriated tungsten tip are separately cooled and the argon gas at a suitable pressure flows between the orifice and tip. It is the design and manufacture of the tip/orifice which controls the argon velocity that is critical to the life of the tip. Argon flows are typically 25-150 l/min/1,000A and torches with power ratings up to 5 MW are being operated. Cathode lives of 30 to 200 hours are being reached depending on the specific operating conditions.

Plasma Arc Melting

The plasma state is generated when a gas is raised to such a temperature that the gas ionises and conducts electrical energy. Any gas can be energised into the plasma state which exists over a wide range of pressure and temperature. The plasmas of use in the steelmaking industries are called thermal plasmas generated by electric arcs at atmospheric pressure generally using argon as the plasma gas. Further, these plasmas are AC or DC and of the transferred arc type.

The arc is struck between the thoriated tungsten cathode and the scrap or liquid steel. The return current path may be another plasma torch, or some solid electrode of graphite, steel or copper. The plasma gas, generally argon, flows at 25-150 l/min/1,000A and carries the energy to the metallurgical system and also protects the electrode from oxidation.

Currently, plasma torches are limited to about 10,000A and 5 MW depending on the voltage that can be generated. Such a torch would be 100-150mm diam compared to a 500-600 mm diam graphite electrode. Graphite electrodes are made up to 1250mm diam and will carry 15 to 20 MW.

The potential advantages of plasma arc melting over graphite electrode arc melting are:

1. No carbon pick-up from electrodes.

2. Higher metallic yield because of the reduced oxidation of light scrap since melting is under an argon atmosphere. Higher retention of high value alloying elements such as Cr, Ti, Mo and V.

3. Significant reduction in physical noise for the operators (less than 90 dB).

4. Significant reduction in "flicker" on the grid system.

As for the DC arc furnace, the body of the plasma arc furnace is like the conventional AC graphite electrode furnace. The lighter mechanical plasma torches can be fitted into the side walls of the furnace thus simplifying roof construction and helping in rapid scrap charging. Because of torch power limitations, furnace power, even with 4 or 6 plasma torches would be limited to 20-30 MW.

The VOEST Alpine Frietal Unit

The first plasma furnace was built by Frietal (East Germany) as a 15 tonne unit later increased to 40 tonnes. Voest Alpine in Austria also installed a 45 tonne unit in their Linz Works in the early 1980 s. The four torch furnace of this type is shown in figure 8.2.

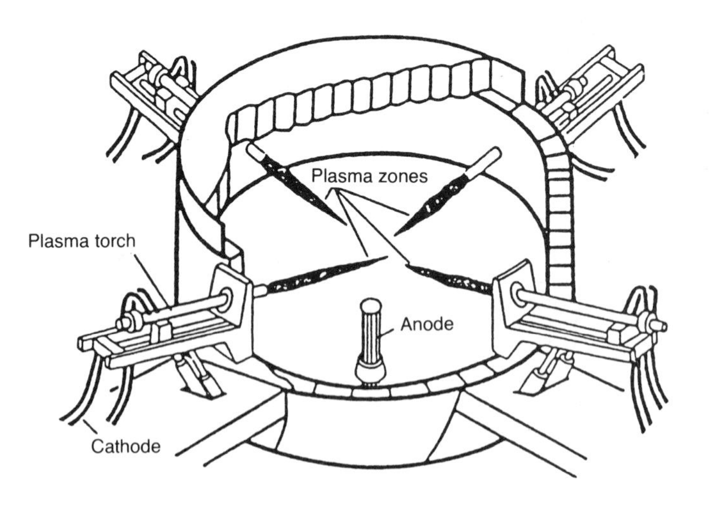

8.2 The VOEST Alpine Frietal plasma furnace

About 85,000 tonnes of carbon and low alloy steel have been made in this 24-30 MW unit. To ensure good electrical contact with the base return electrode, a molten heel of up to 5 tonnes is left in the furnace. Alternatively, oxy-fuel burners are used after charging to knock down the scrap and provide the heel. This practice has the further advantage of reducing scrap volume and reducing power to the melt. 13 MW are used for scrap densities of 1.5 t/m^3 compared with 10 MW for 2 t/m^3.

Typical operating data for the furnace are given in Table 8.1

TABLE 8.1

Published operating data on the

Voest Alpine DC plasma furnace

Shell diam	5.8m
Capacity	45 t
Torches	4
Power	24-30 MW
Melting rate	25-30t/h
Melt down energy	450 kW h/t
Tap energy	520-620 kW h/t
Argon use	2-3 m^3/t
Oxygen use	7-11 m^3/t
Yield increase	1.5%
Torch life	
(Kg of graphite equivalent)	2.5-3 kg/t
Noise (max)	86dBA

The Krupp AC Plasma System

After pilot work on a 1.5 MW unit, a 20 MW, 10 tonne furnace for scrap melting was installed in the Siegen works of Krupp Stahl in Germany. Being AC, the unit has three torches which

are installed through the roof as in a conventional arc furnace. One advantage is that there is no base return electrode but the torches generate more heat at the electrode tip than in a DC cathodic torch. AC arcs repel one another so the three torches can be placed relatively close to one another and centrally to reduce the heat load on the furnace walls. The published operating data is given below in Table 8.2.

TABLE 8.2

Operating data on the Krupp AC plasma furnace

Capacity	10 t
Torches	3
Power	3.6 MW
Argon use	9 m^3/t
Torch life (kg of graphite equivalent)	1.5 kg/t
Noise	88-92 dBA

These two units have clearly shown that plasma arc can effectively melt and make steel. Plasma units have, however, need of development to increase capacity and melting rates that can now be reached with conventional graphite electrode arc furnaces. Typical EAF capacity today is 50-150t with units up to 200 tonnes. Production rates of 40-60 tonnes/hour are typical. These values are unlikely to be matched by plasma within the next decade but there is possibly a place for plasma in the smaller capacity, high alloy steel manufacture, where production rate is of less concern than the retention of alloys such as Cr, Ti, Mo and V.

SECONDARY PLASMA PROCESSING

Currently about 50% of world steel production is continuously cast and by the year 2000, it is very likely that 80-90% will be processed by this route. Consistent, steady and controllable steel temperature into the casting mould gives reliable caster performance and good consistent metallurgical quality to the billet, bloom or slab. Further, the nearer the mould entry temperature is to the liquidus of the steel being cast, the finer and more equiaxed is the grain structure of the steel. With the temperature variations inherent in the steelmaking cycle

and the decrease in temperature from the furnace, to ladle, to tundish and into mould, a significant drop has to be allowed. Cold tapping, cold ladles and delays mean that there has to be a significant margin over liquidus when the furnace is tapped. Depending on furnace capacity (the lower the volume, the greater the heat loss) the specific site layout and sections being cast, tapping at up to 200°C above liquidus is needed. Reducing this by 20-30°C can bring about significant savings in power (for the EAF) and refractories and increase productivity by reducing tap to tap time. Such reductions are possible by having secondary heating at the ladle or tundish.

Ladle Heating and Refining

This topic has been dealt with in Chapter 7. In most cases, the ladle heating system is an AC three phase system using graphite electrodes and a conducting slag. No major modifications are required on the ladle as is the case if induction heating is used. Induction requires either stainless steel ladle shells or "loops" through which steel can flow and be inductively heated. Because of these complications, electrode heating is the system most often chosen. Both systems will reheat the steel during refining, alloying, stirring or operational delays and permit a reduction in furnace tapping temperature.

Electrical capacities up to 30 MW are used and the use of plasma is being evaluated for this application. Compared to carbon arc heating, plasma has some unique advantages.

1. Higher efficiency of heat transfer from the argon stabilised plasma arc (70-80%).

2. Better retention of the argon atmosphere over the slag/metal due to the smaller diameter of the plasma torch, its cold outer surface and the stable position during operation.

The technique was evaluated at the US Steels, Lorraine Works on a 220 tonne ladle using a 4 MW DC torch with a graphite return electrode. Chaparral Steel, Texas also has a 4 MW system on a 150 tonne ladle where the return current path is through conductive refractory bricks in the ladle wall. This latter system reduces the level of carbon contamination of the steel. The Lorraine works system and operating costs are

shown below in figure 8.3 and Table 8.3.

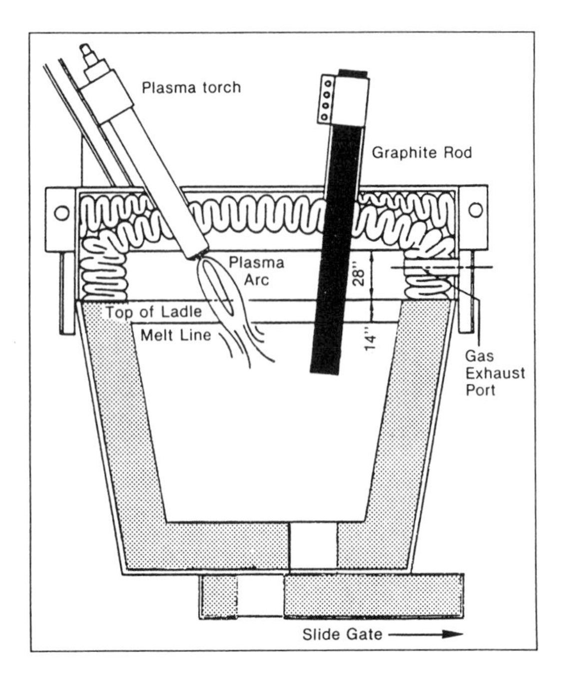

8.3 The Lorraine Works ladle heating system

An AC arc plasma system is being used on a 30 tonne ladle unit at the Krupp Siegen Works using 20 MVA. AC systems are self sustaining and need no electrical connections other than the three AC plasma torches.

Tundish Heating

Control of temperature at the mould entry point should give more accurate and consistent control than in the ladle which supplies the tundish. At the tundish, the ladle is no longer heated and for long hold times will provide steel at predictable but progressively lower temperatures. At ladle changeover

TABLE 8.3

Published cost breakdown and basis of breakdown of the Lorraine system

Breakdown

Direct cost ($)	
Electric power	0.73
Plasma torch maintenance	0.20
Labour, operating and maintenance	0.22
Refractory	0.35
Indirect Cost ($)	
Capital equipment	0.27
Total cost/ton	1.97

Basis

1. Ladle size: 220 tons (200/t)
2. Ladle treatment time at the PLH : 1h
3. PLH* treats one ladle every 2 h
4. Capital cost for the PLH equipment : $1.5 million
5. Power cost : 4 cents/kWh
6. Plasma gas cost (nitrogen) : $4 per 1,000 ft^3 ($0.14/m^3).

* PLH = plasma ladle heater.

during link or sequence casting there can be significant change from the cooler steel in the tundish to hotter steel from a new ladle. Such changes can be overcome to give a ±5°C control of mould temperature which can be closer to the liquidus of the steel being cast. Induction tundish heaters are being evaluated but some plasma systems are installed in operational tundishes. These are given below in Table 8.4.

TABLE 8.4

Plasma tundish heating systems

Works	Type	Capacity kW
Chaparral, USA	DC	300
Deltasider, Italy	AC	2,000
Kobe	AC	2,000
Nippon Steel, Japan	DC	1,000
Aichi, Japan	DC	300
NKK, Japan	DC	1,400

The AC system has three torches and no need of a mechanical return electrode in the tundish. The fact that three torches have to be incorporated into the tundish may be a disadvantage.

The DC system usually has one cathodic torch and a mechanical carbon or steel return electrode built into the end wall of the tundish. Recent development of an anodic DC torch now means that such mechanical electrodes are not obligatory. The plasma torch can be positioned where a particular tundish needs the heating.

The Nippon Steel unit using a Tetronics torch on a 14 tonne slab caster is shown in figure 8.4.

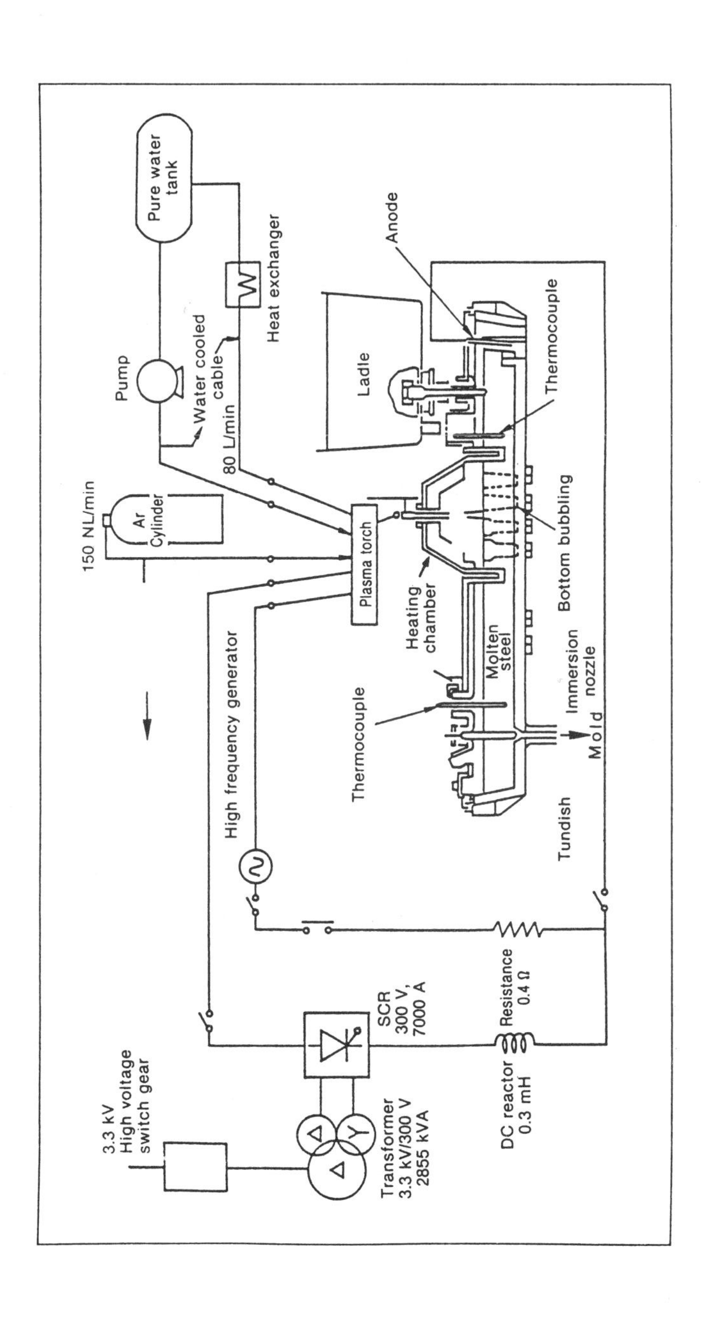

8.4 The Nippon Steel/Tetronics tundish heating system

The one MW plasma torch operates at 70-80% efficiency thus requiring about 15 kWh/°C/tonne/min. The variation of the temperature in a cast can be held within ± 5°C even during ladle changeover. The tap temperature of the 100 tonne BOF which provides the steel can be reduced by about 20°C which offers significant savings in production time and refractories in the BOF.

Plasma Dust Processing

One unavoidable consequence of steelmaking is the production of very fine dust of 2μm or less. The dust can be as much as 2% of the charge weight and from an arc furnace will contain Zn, Pb and Cd which can be leachable from a dust dump. From the AOD, the dust contains valuable Ni, Cr and Mo as well as Pb and Zn. Environmental legislation in USA is beginning to require that these dusts are rendered non-hazardous on the producing site and it is likely that the EC will bring in some form of similar controlling legislation soon.

TABLE 8.5

Scandust iron composition

%	Carbon steel dust	Alloy dust
C	4 - 6	4 - 6
Si	0.2 - 1.0	0.1 - 1.5
Mn	0.1 - 3.0	2 - 5
P	0.15 - 0.45	0.05 - 0.08
S	0.03 - 0.15	0.03 - 0.15
Cu	0.1 - 1.0	0.1 - 1.0
Cr		16 - 18
Ni		6 - 8
Mo		0.5 - 2.0
Fe	90 - 95	60 - 70

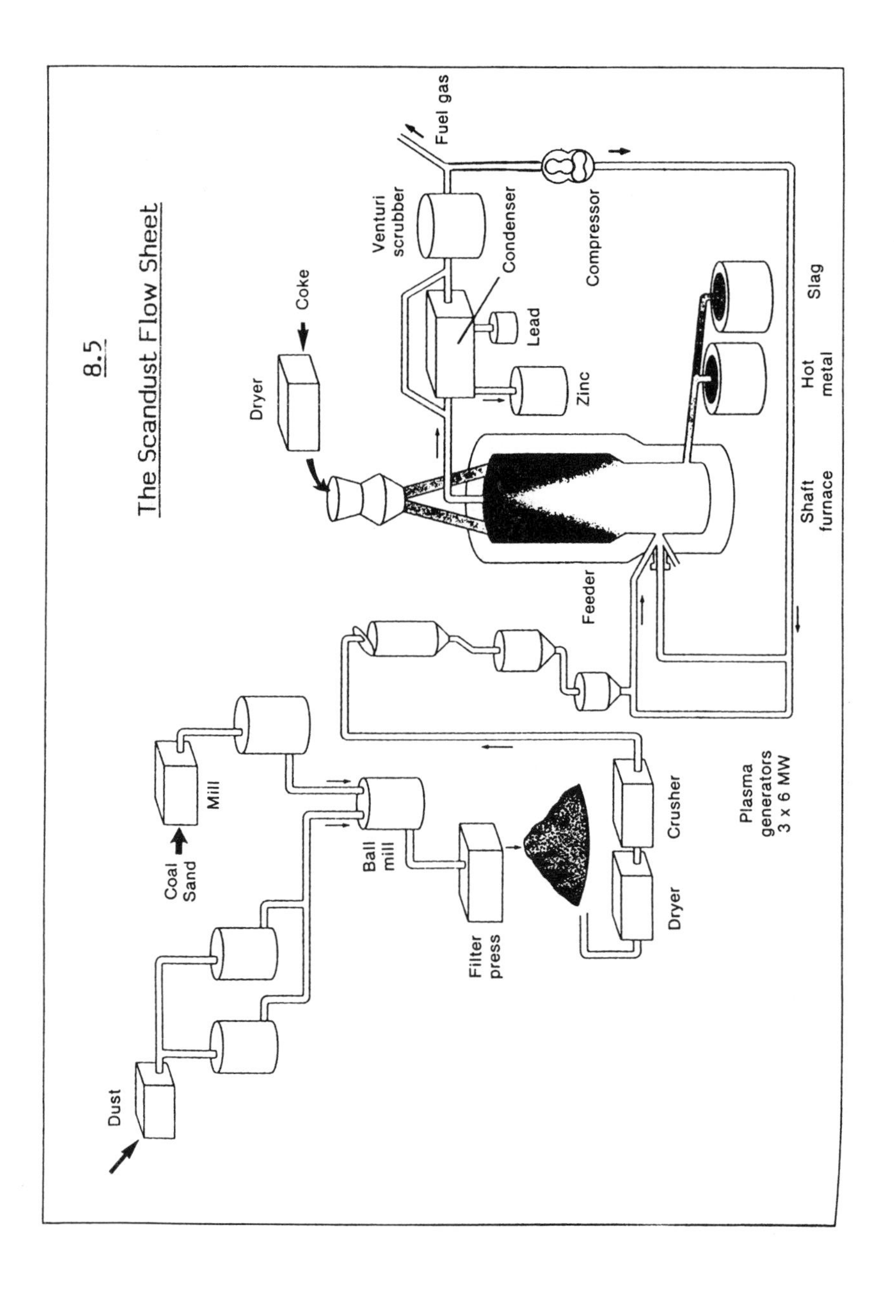
8.5
The Scandust Flow Sheet
Dust
Coal
Sand
Mill
Ball
mill
Filter
press
Dryer
Crusher
Plasma
generators
3 x 6 MW
Feeder
Dryer
Coke
Shaft
furnace
Hot
metal
Slag
Zinc
Lead
Condenser
Venturi
scrubber
Compressor
Fuel gas

The Scandust Plant in Sweden is designed to treat 70,000 tonnes of dust per year using three SKF plasma gas heater torches of 18 MW total capacity. The flow sheet is shown in figure 8.5.

The unit is fed by dusts from several plants and runs campaigns on carbon steel and stainless steel dusts, the major operation has been on the latter. Dusts, oxide waste, slag and coal are injected into the shaft furnace. Prime Western Grade Zn (98% Zn) is recovered from EAF dust together with a low alloy iron. From stainless steel dusts an alloy iron is recovered. The composition of these irons is given in Table 8.5 from published data.

Tetronics plasma dust plants are being built in the UK and USA to treat stainless and carbon dusts. These are small plants designed to treat the dust arisings on site in line with the steel furnace dust system using a smelting reactor. Zn can be recovered by a condenser and from carbon steels, the other "product" is a dumpable non-leachable slag. For stainless steel dusts a Ni, Cr, Mo, Fe alloy is produced for return within the works. The composition of stainless dust and plasma produced alloy from it are given in Table 8.6.

TABLE 8.6

Dust and alloy analysis from Tetronics plasma reactor

Dust		Alloy	
Compound*	%	%	Metal
Fe_2O_3	42	60-65.0	Fe
Cr_2O_3	16	21.0	Cr
NiO	3	5.4	Ni
MnO	6	3.6	Mn
MoO_3	1	1.6	Mo
SiO_2	7	0.7	Si
Other oxides	25		
	-	5.0	Carbon
		0.08	S
		0.06	P

*The recovery of the Ni, Cr, Mo is of the order of 95%.

PLASMA DEVELOPMENTS

It has to be emphasised that the plasma processes have only been used in operational steelworks for a few years. The initial growth of plasma techniques is probably into tundish heating where powers of 2 MW are adequate. Development of 10-20 MW torches will undoubtedly see the true plasma furnace being used to melt steels in larger (say 100 tonnes) furnaces.

Its use in individual steelworks as a means of processing hazardous dusts into valuable products and non water leachable materials, is likely to grow at an even higher rate as the need to recycle raw materials and environmental considerations become even more pressing.